Dave McDonald

Hamster S.A.M.
Odd-ventures in Space!

SWEET
CORN

Library of Congress Control Number: 2013936349
ISBN: 978-0-9798445-2-2 (paperback)

writer and artist
Dave McDonald

editor
Janna Morishima

Acknowledgements: This book would not have been possible without the
thoughtful guidance of Janna Morishima. Her enthusiastic spirit, keen
insight and generosity are deeply appreciated. Thank you so much, Janna.
I am grateful to cartoonist and friend Steve Smith, who arranged for Hamster Sam to meet "Astronaut Greg". And of course, my sincere thanks to
NASA astronaut Greg "Box" Johnson for making this project such a thrill.

NASA photographic images used in this publication appear
courtesy of and are copyright 2013 by NASA and by
the Space Telescope Science Institute (STScI).

To learn more about Dave and Hamster Sam, please visit them at:
www.davemcdonald.com

Printed in the United States of America.

*
First Edition

For Jacala Richardson and for every teacher
who has encouraged somebody's dream.

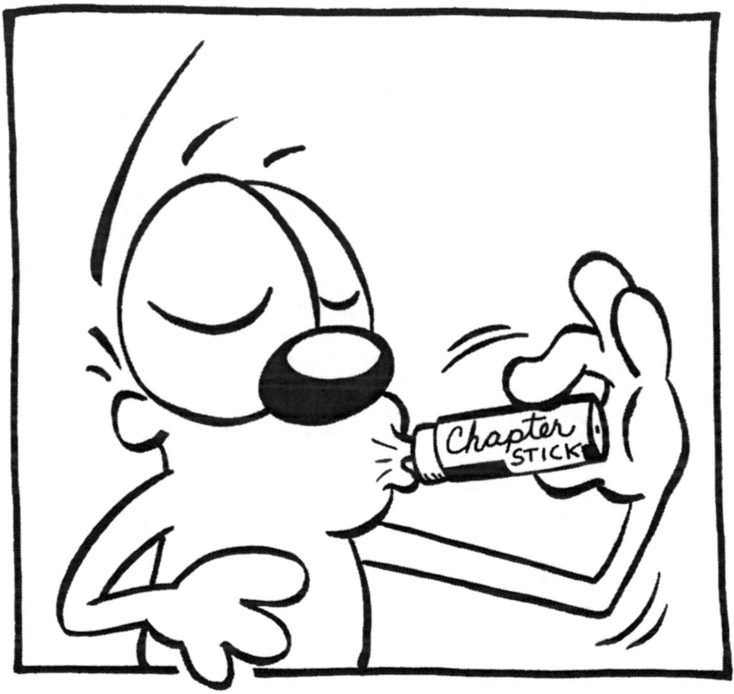
Chapter
STICK

Chapters

STURGEON GENERAL'S
WARNING

WARNING! Contains side-splitting humor. May cause rolling fits of laughter, dislocated funny bones and quite possibly initiate a life-long enjoyment of reading. Turn page at your own risk!

Chapter One

The Name's Sam...
Hamster Sam!

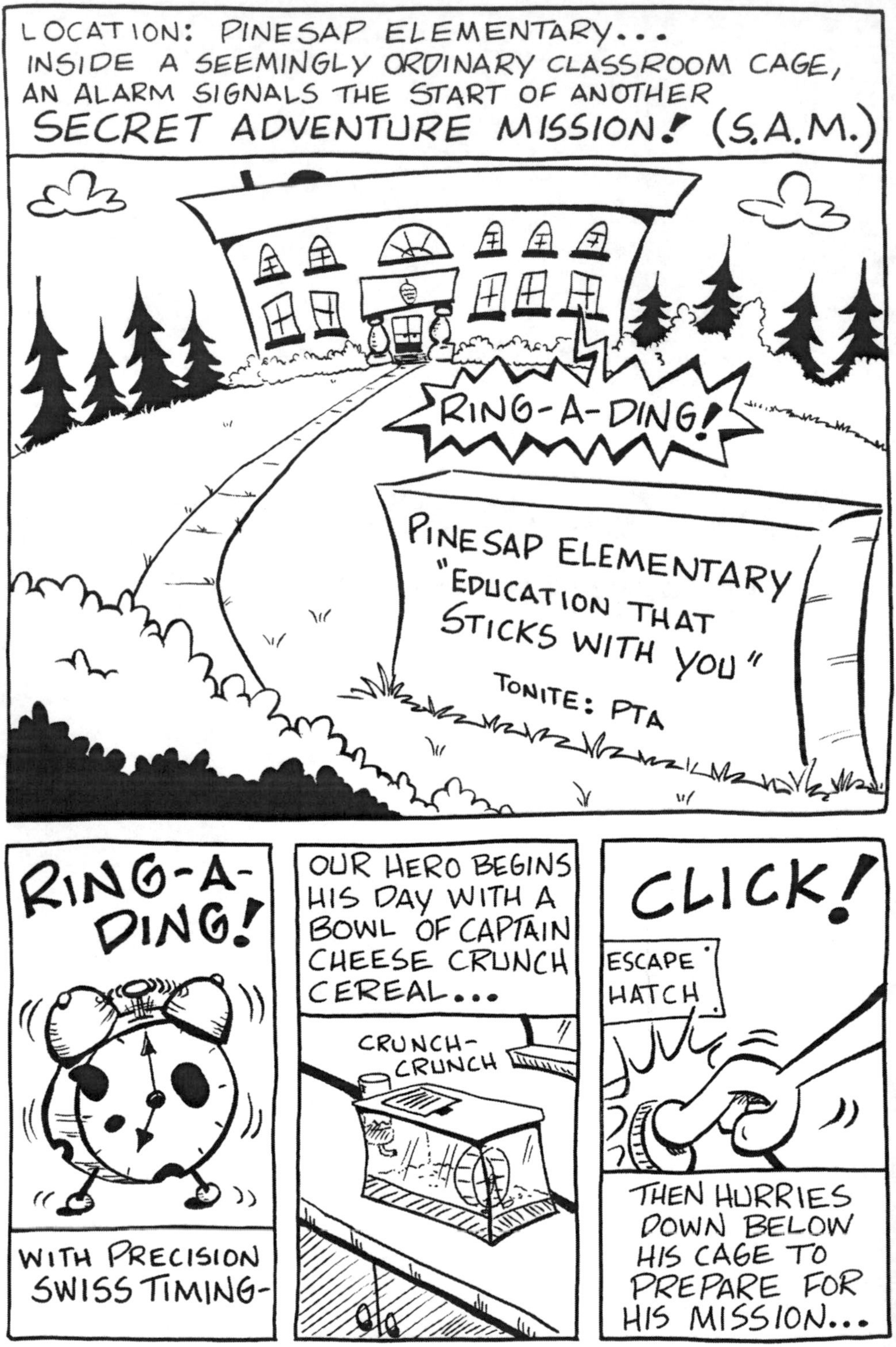
LOCATION: PINESAP ELEMENTARY...
INSIDE A SEEMINGLY ORDINARY CLASSROOM CAGE, AN ALARM SIGNALS THE START OF ANOTHER
SECRET ADVENTURE MISSION! (S.A.M.)
RING-A-DING!
PINESAP ELEMENTARY "EDUCATION THAT STICKS WITH YOU"
TONITE: PTA
RING-A-DING!
WITH PRECISION SWISS TIMING-
OUR HERO BEGINS HIS DAY WITH A BOWL OF CAPTAIN CHEESE CRUNCH CEREAL...
CRUNCH-CRUNCH
CLICK!
ESCAPE HATCH
THEN HURRIES DOWN BELOW HIS CAGE TO PREPARE FOR HIS MISSION...

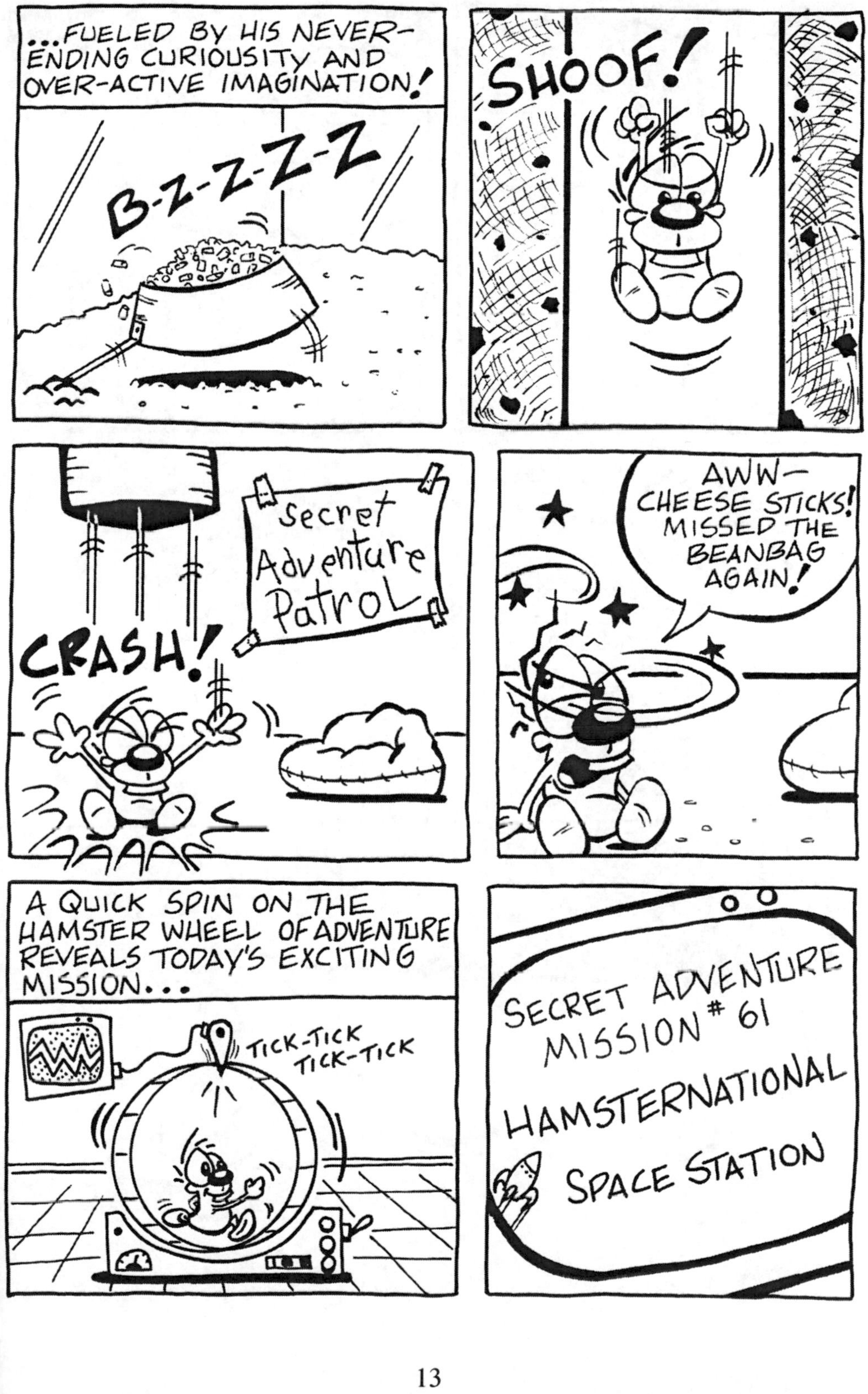
...FUELED BY HIS NEVER-ENDING CURIOUSITY AND OVER-ACTIVE IMAGINATION!
B-Z-Z-Z-Z-Z
SHOOF!
Secret Adventure Patrol
CRASH!
AWW-CHEESE STICKS! MISSED THE BEANBAG AGAIN!
A QUICK SPIN ON THE HAMSTER WHEEL OF ADVENTURE REVEALS TODAY'S EXCITING MISSION...
TICK-TICK TICK-TICK
SECRET ADVENTURE MISSION # 61
HAMSTERNATIONAL
SPACE STATION

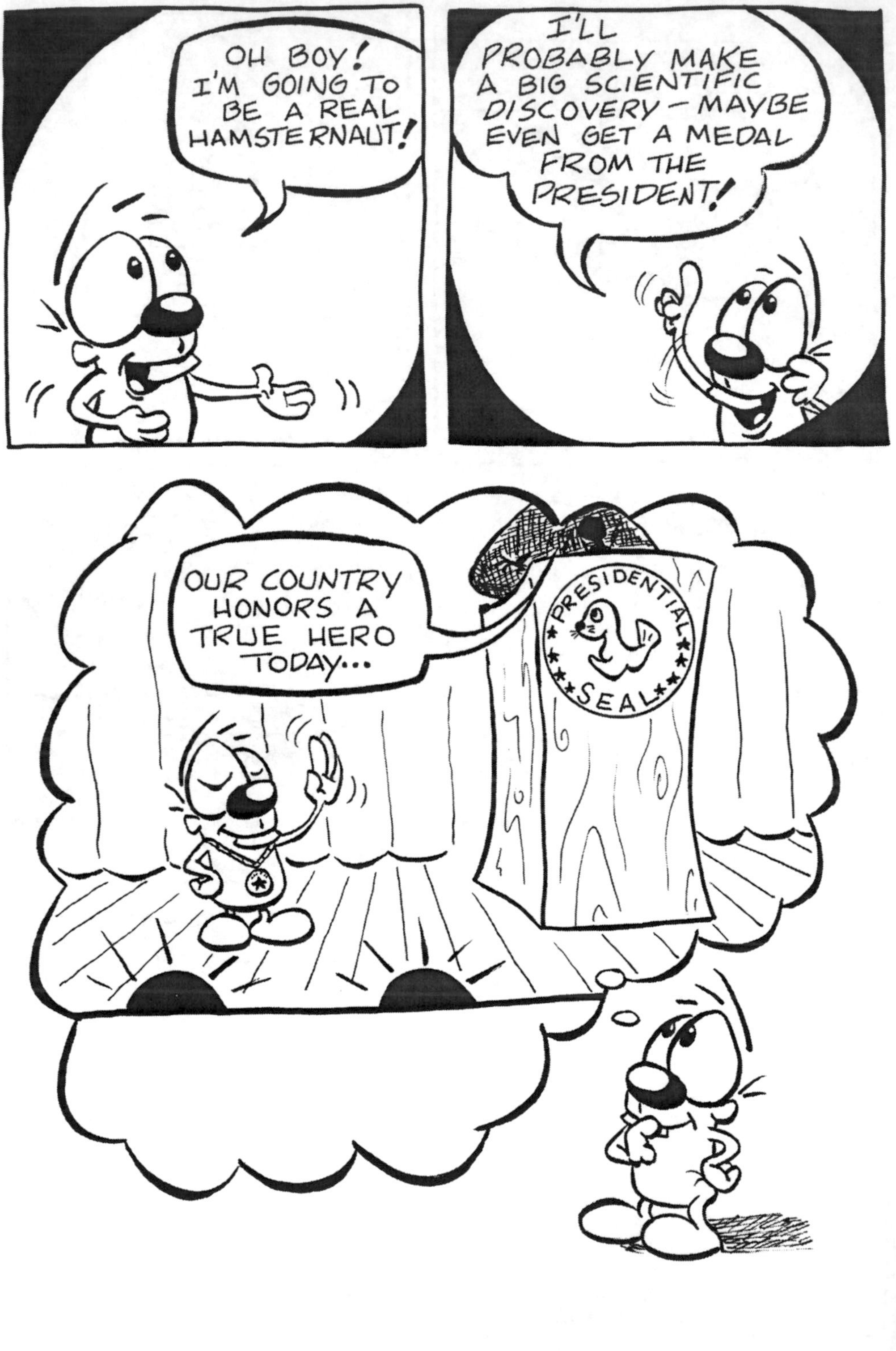
OH BOY! I'M GOING TO BE A REAL HAMSTERNAUT!
I'LL PROBABLY MAKE A BIG SCIENTIFIC DISCOVERY – MAYBE EVEN GET A MEDAL FROM THE PRESIDENT!
OUR COUNTRY HONORS A TRUE HERO TODAY...
PRESIDENTIAL SEAL

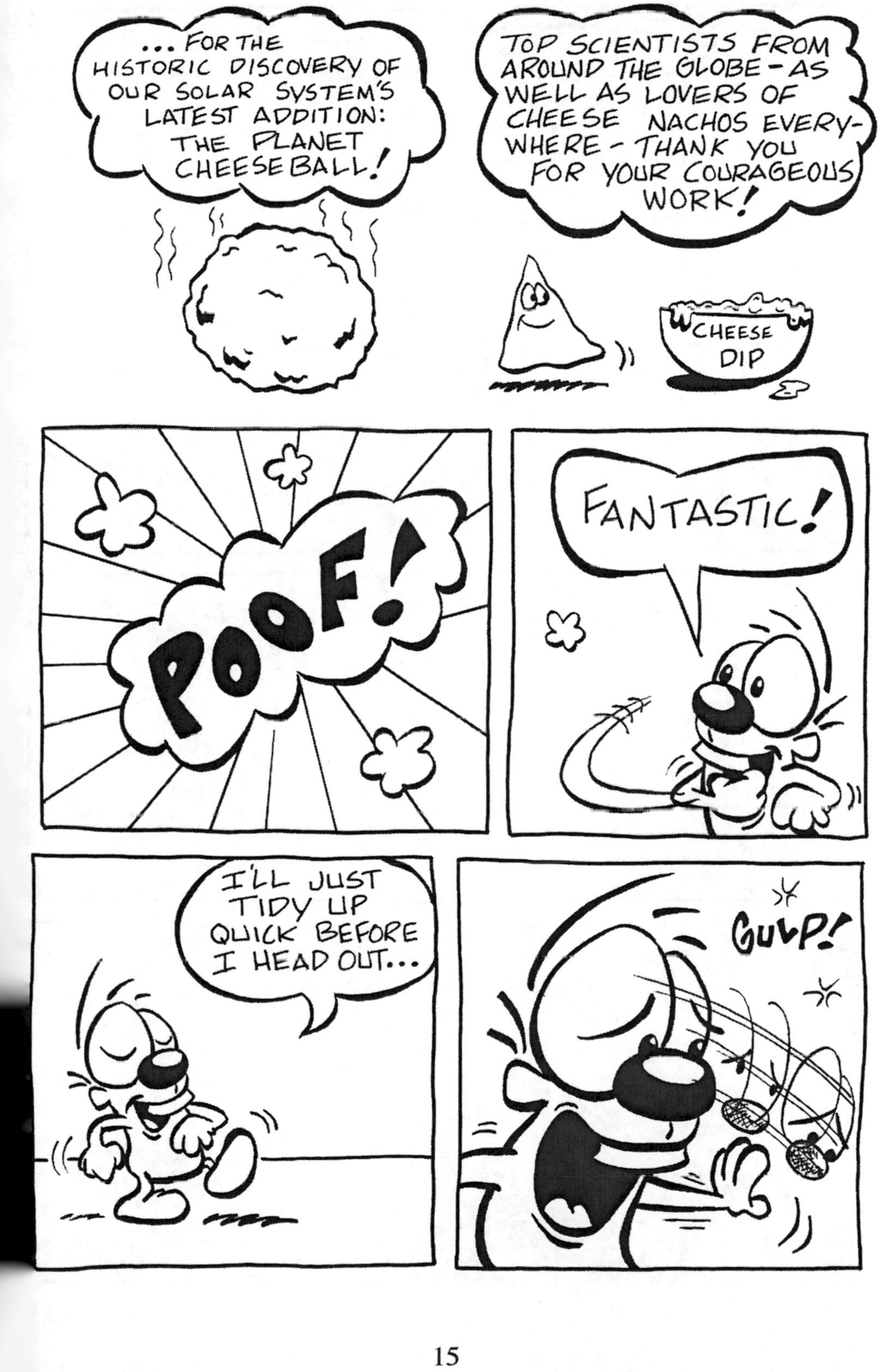
... FOR THE HISTORIC DISCOVERY OF OUR SOLAR SYSTEM'S LATEST ADDITION: THE PLANET CHEESEBALL!
TOP SCIENTISTS FROM AROUND THE GLOBE - AS WELL AS LOVERS OF CHEESE NACHOS EVERY-WHERE - THANK YOU FOR YOUR COURAGEOUS WORK!
CHEESE DIP
POOF!
FANTASTIC!
I'LL JUST TIDY UP QUICK BEFORE I HEAD OUT...
GULP!

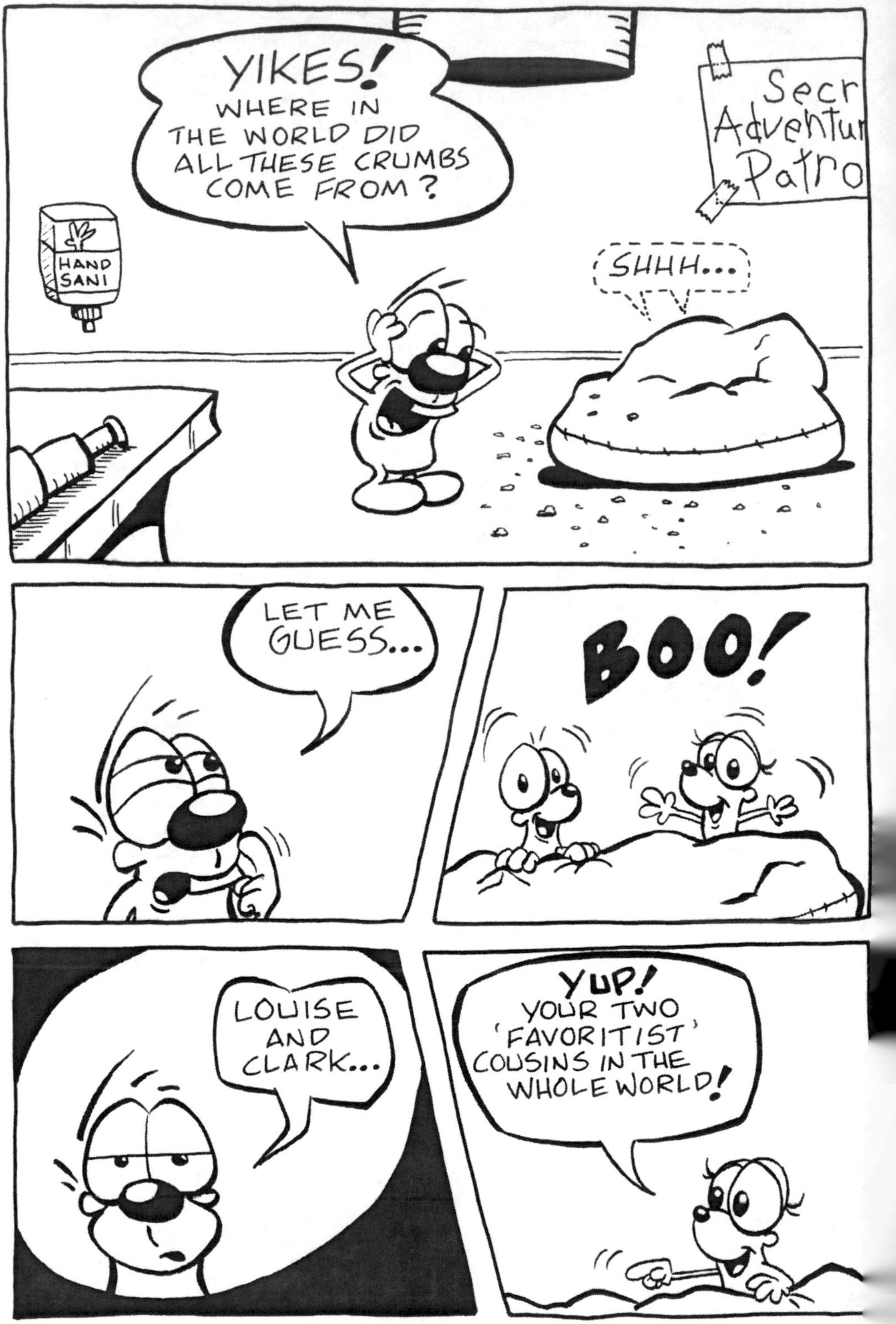
YIKES! WHERE IN THE WORLD DID ALL THESE CRUMBS COME FROM?
HAND SANI
SHHH...
LET ME GUESS...
BOO!
LOUISE AND CLARK...
YUP! YOUR TWO 'FAVORITIST' COUSINS IN THE WHOLE WORLD!

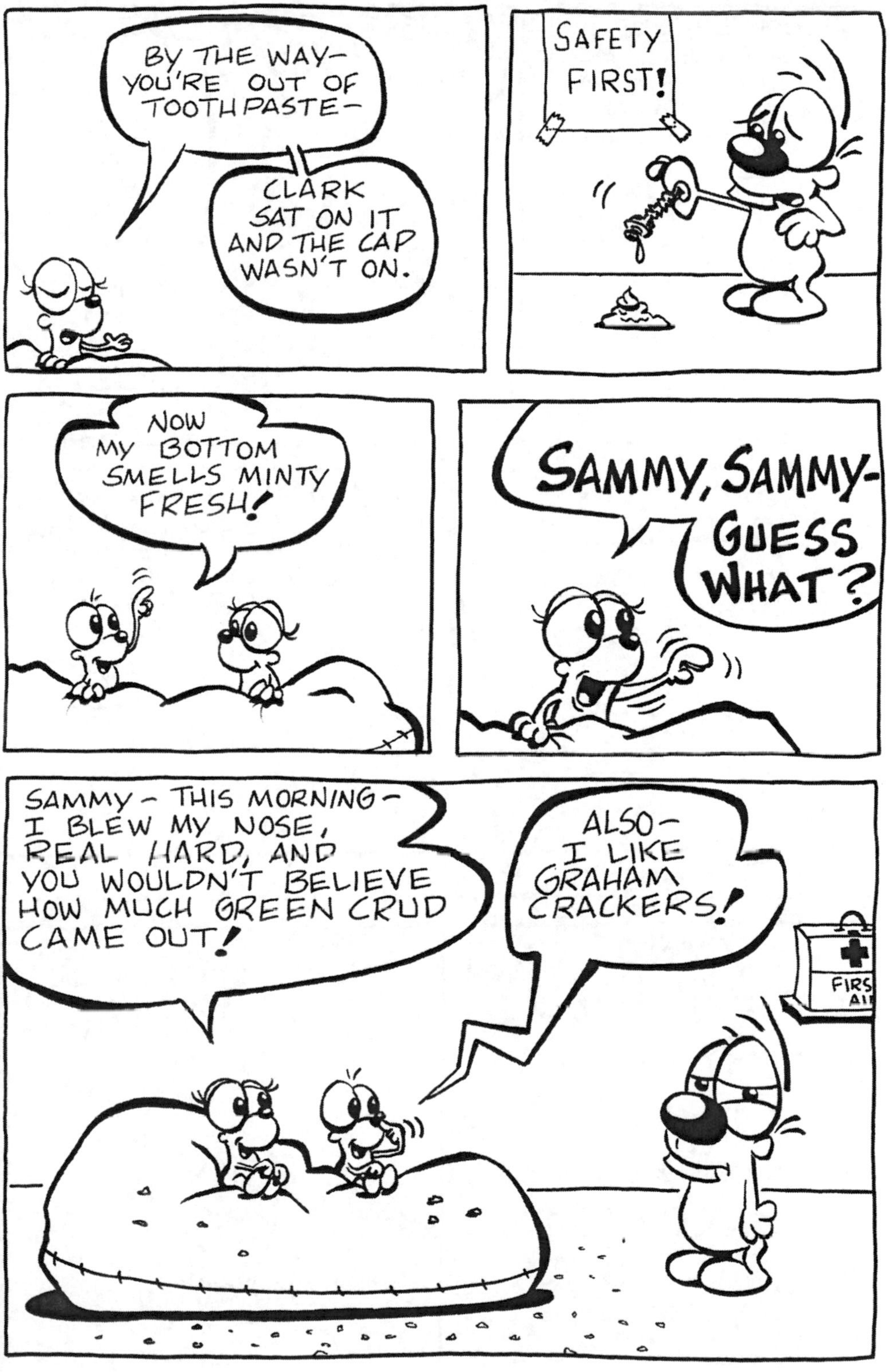
BY THE WAY— YOU'RE OUT OF TOOTHPASTE—
CLARK SAT ON IT AND THE CAP WASN'T ON.
SAFETY FIRST!
NOW MY BOTTOM SMELLS MINTY FRESH!
SAMMY, SAMMY- GUESS WHAT?
SAMMY - THIS MORNING - I BLEW MY NOSE, REAL HARD, AND YOU WOULDN'T BELIEVE HOW MUCH GREEN CRUD CAME OUT!
ALSO- I LIKE GRAHAM CRACKERS!

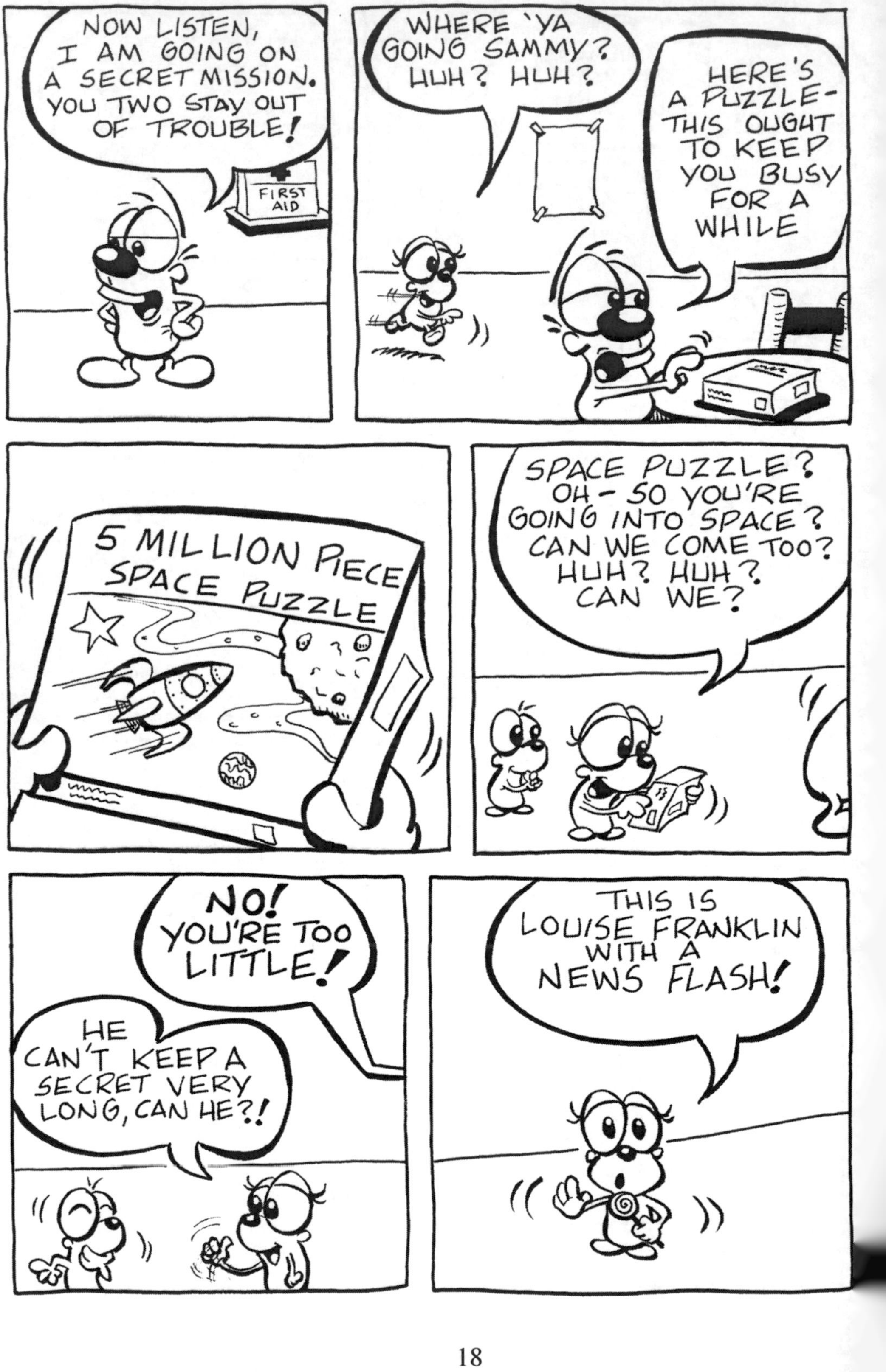
NOW LISTEN, I AM GOING ON A SECRET MISSION. YOU TWO STAY OUT OF TROUBLE!
FIRST AID
WHERE 'YA GOING SAMMY? HUH? HUH?
HERE'S A PUZZLE- THIS OUGHT TO KEEP YOU BUSY FOR A WHILE
5 MILLION PIECE SPACE PUZZLE
SPACE PUZZLE? OH - SO YOU'RE GOING INTO SPACE? CAN WE COME TOO? HUH? HUH? CAN WE?
NO! YOU'RE TOO LITTLE!
HE CAN'T KEEP A SECRET VERY LONG, CAN HE?!
THIS IS LOUISE FRANKLIN WITH A NEWS FLASH!

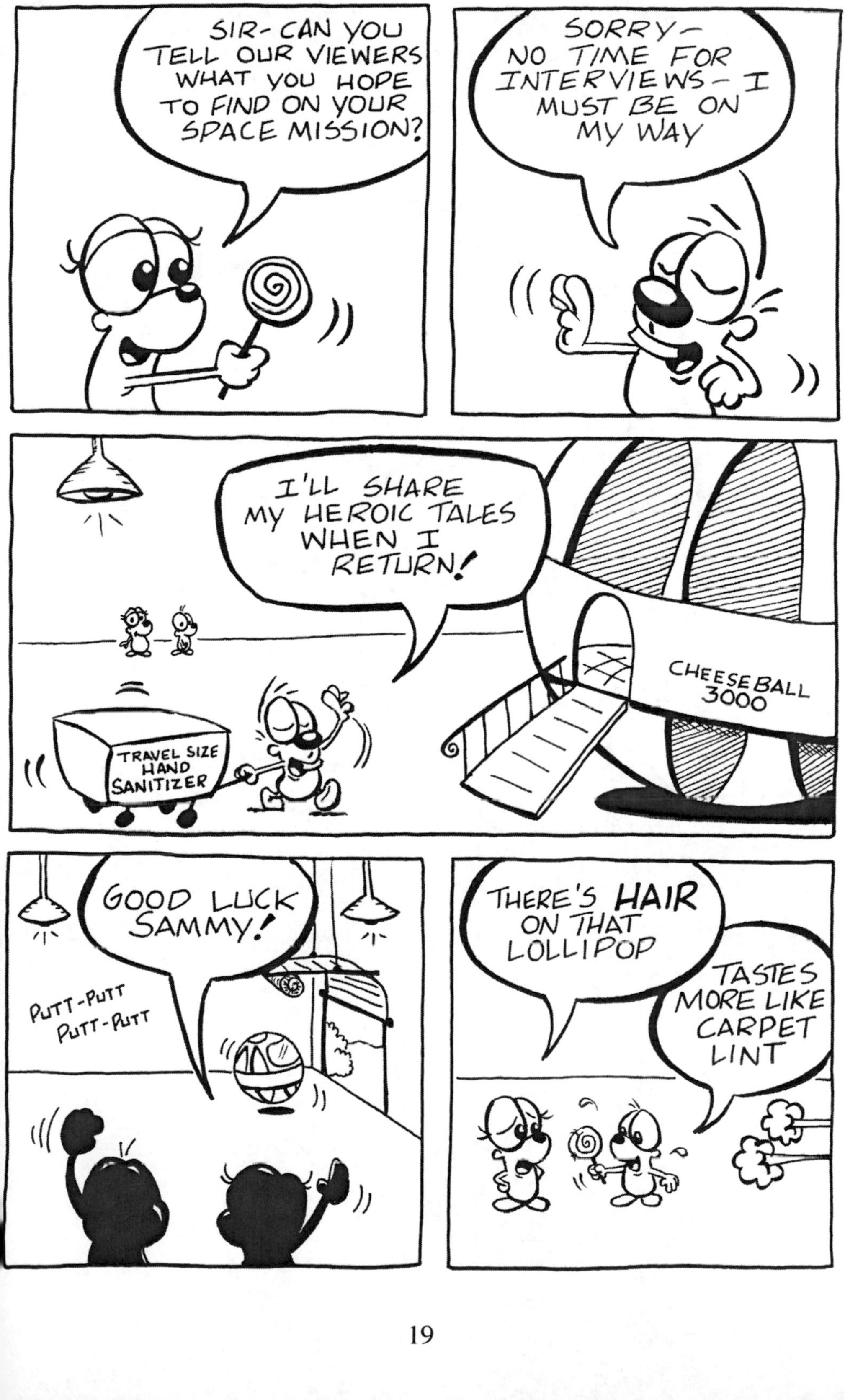
SIR- CAN YOU TELL OUR VIEWERS WHAT YOU HOPE TO FIND ON YOUR SPACE MISSION?
SORRY- NO TIME FOR INTERVIEWS- I MUST BE ON MY WAY
I'LL SHARE MY HEROIC TALES WHEN I RETURN!
TRAVEL SIZE HAND SANITIZER
CHEESEBALL 3000
GOOD LUCK SAMMY!
PUTT-PUTT PUTT-PUTT
THERE'S **HAIR** ON THAT LOLLIPOP
TASTES MORE LIKE CARPET LINT

Chapter Two

Mullet Over

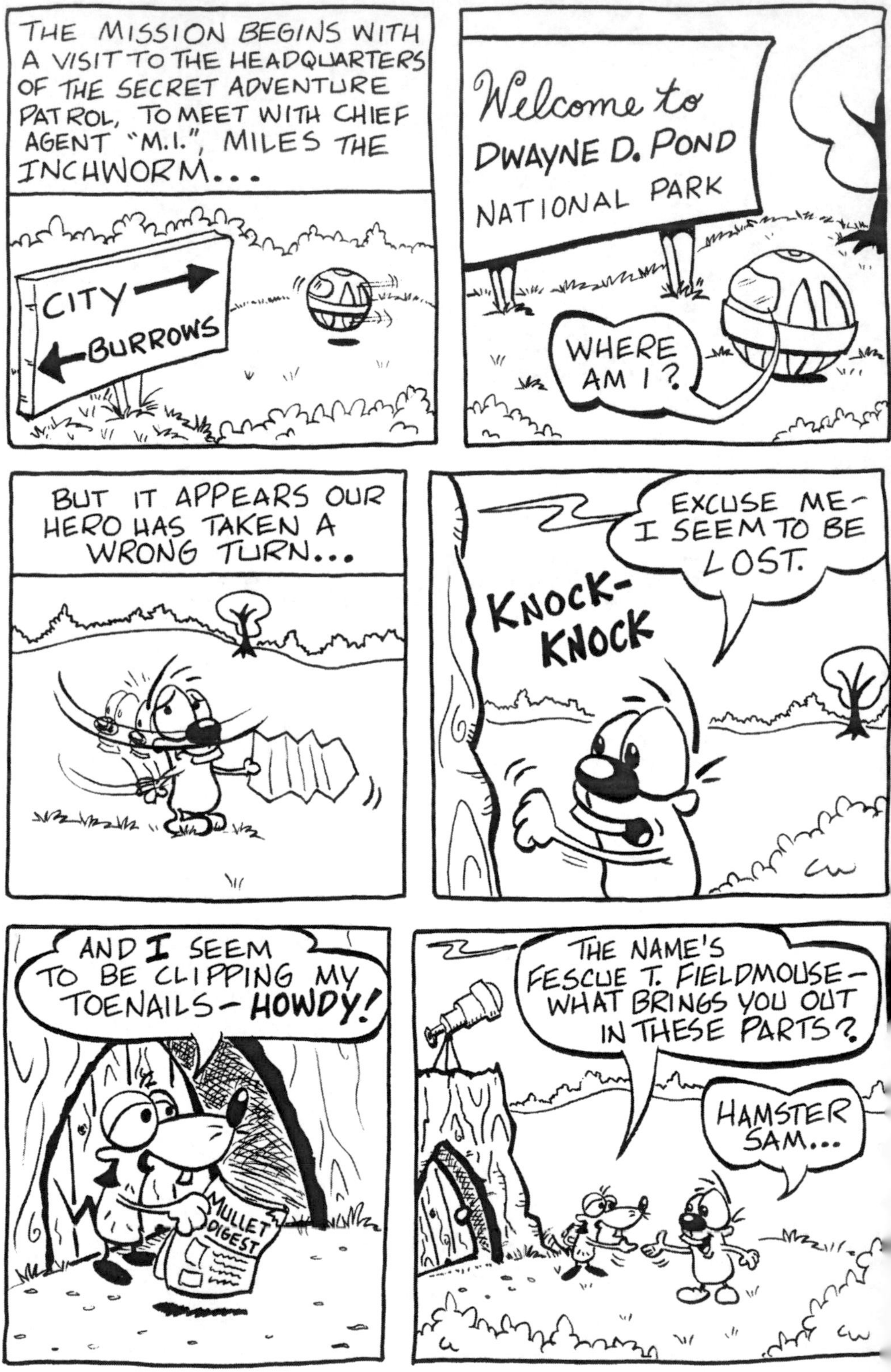
THE MISSION BEGINS WITH A VISIT TO THE HEADQUARTERS OF THE SECRET ADVENTURE PATROL, TO MEET WITH CHIEF AGENT "M.I.", MILES THE INCHWORM...
CITY
BURROWS
Welcome to DWAYNE D. POND NATIONAL PARK
WHERE AM I?
BUT IT APPEARS OUR HERO HAS TAKEN A WRONG TURN...
EXCUSE ME- I SEEM TO BE LOST.
KNOCK-KNOCK
AND I SEEM TO BE CLIPPING MY TOENAILS- HOWDY!
MULLET DIGEST
THE NAME'S FESCUE T. FIELDMOUSE- WHAT BRINGS YOU OUT IN THESE PARTS?
HAMSTER SAM...

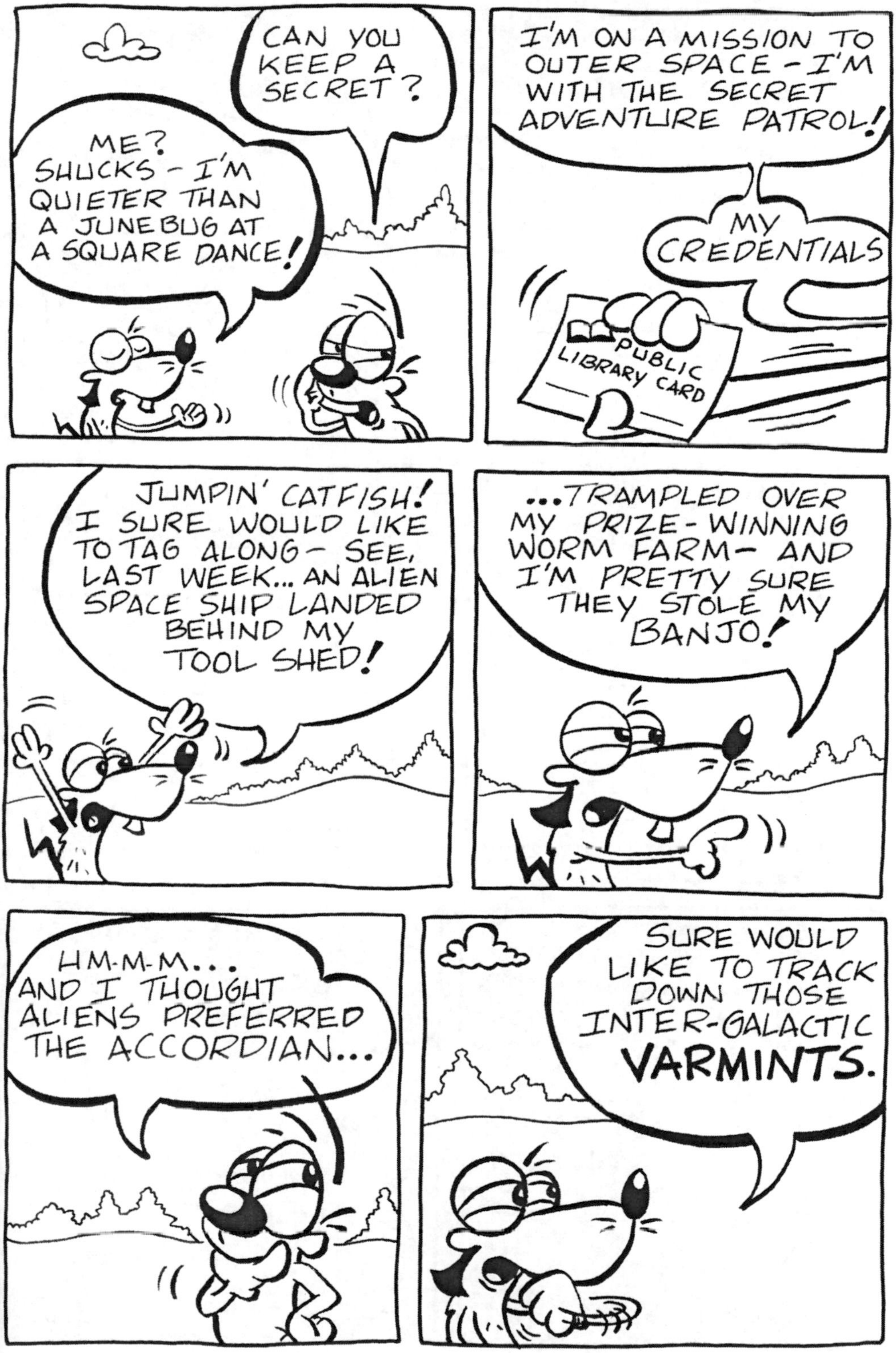
CAN YOU KEEP A SECRET?
ME? SHUCKS - I'M QUIETER THAN A JUNEBUG AT A SQUARE DANCE!
I'M ON A MISSION TO OUTER SPACE - I'M WITH THE SECRET ADVENTURE PATROL!
MY CREDENTIALS
PUBLIC LIBRARY CARD
JUMPIN' CATFISH! I SURE WOULD LIKE TO TAG ALONG- SEE, LAST WEEK... AN ALIEN SPACE SHIP LANDED BEHIND MY TOOL SHED!
...TRAMPLED OVER MY PRIZE-WINNING WORM FARM- AND I'M PRETTY SURE THEY STOLE MY BANJO!
HM-M-M... AND I THOUGHT ALIENS PREFERRED THE ACCORDIAN...
SURE WOULD LIKE TO TRACK DOWN THOSE INTER-GALACTIC VARMINTS.

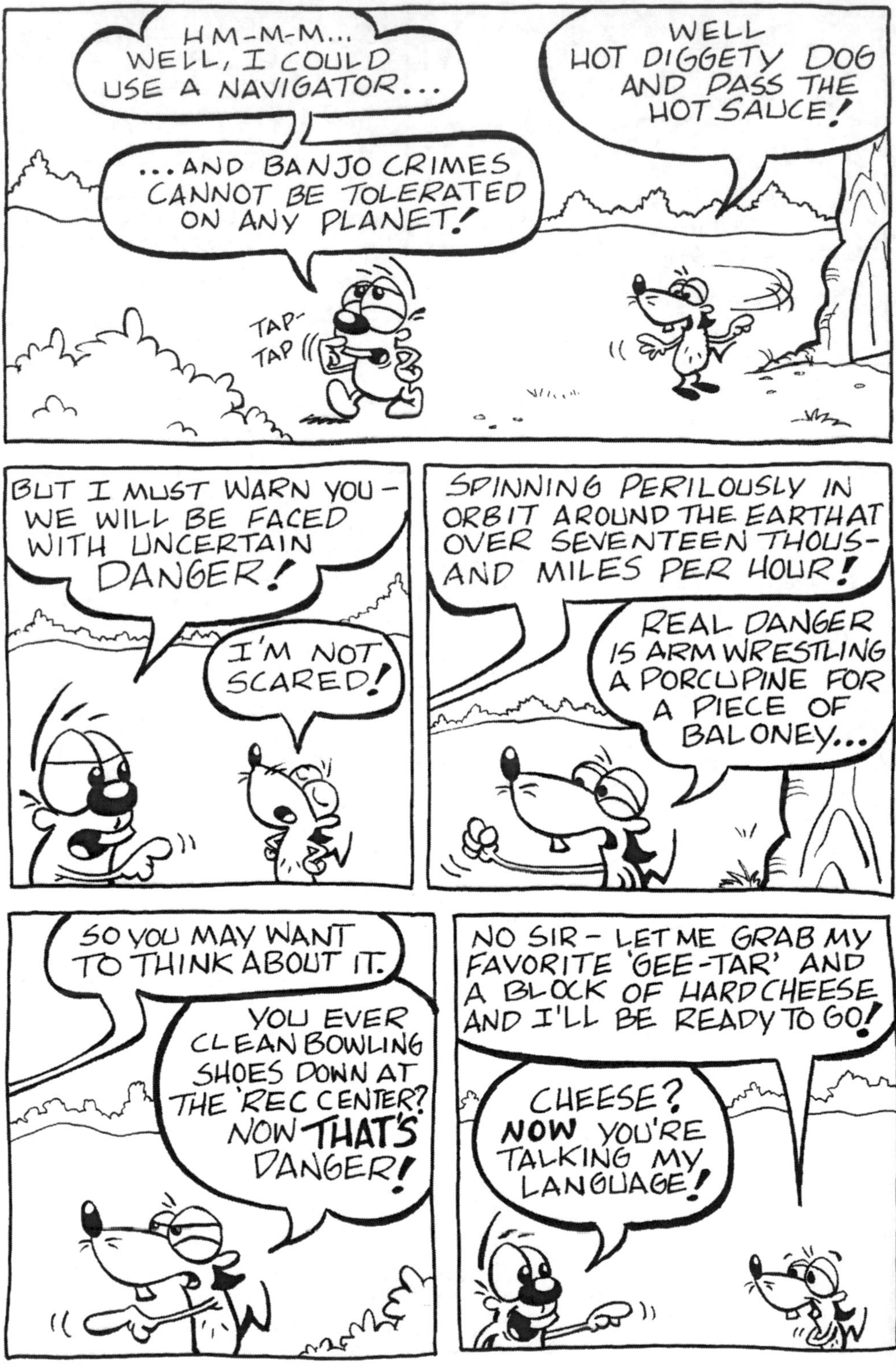
HM-M-M... WELL, I COULD USE A NAVIGATOR...
...AND BANJO CRIMES CANNOT BE TOLERATED ON ANY PLANET!
TAP-TAP
WELL HOT DIGGETY DOG AND PASS THE HOT SAUCE!
BUT I MUST WARN YOU - WE WILL BE FACED WITH UNCERTAIN DANGER!
I'M NOT SCARED!
SPINNING PERILOUSLY IN ORBIT AROUND THE EARTH AT OVER SEVENTEEN THOUS-AND MILES PER HOUR!
REAL DANGER IS ARM WRESTLING A PORCUPINE FOR A PIECE OF BALONEY...
SO YOU MAY WANT TO THINK ABOUT IT.
YOU EVER CLEAN BOWLING SHOES DOWN AT THE 'REC CENTER? NOW THAT'S DANGER!
NO SIR - LET ME GRAB MY FAVORITE 'GEE-TAR' AND A BLOCK OF HARD CHEESE AND I'LL BE READY TO GO!
CHEESE? NOW YOU'RE TALKING MY LANGUAGE!

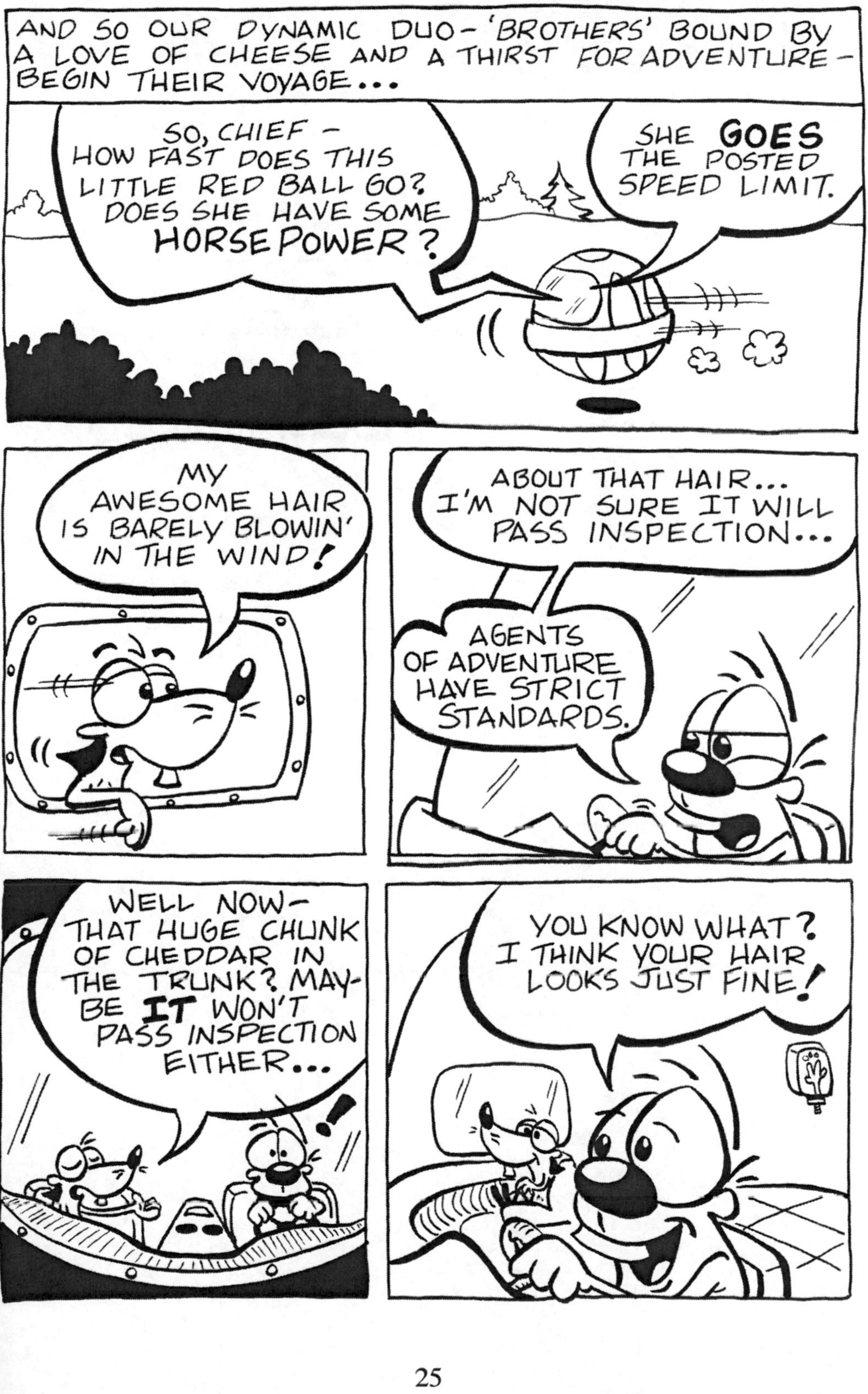
AND SO OUR DYNAMIC DUO- 'BROTHERS' BOUND BY A LOVE OF CHEESE AND A THIRST FOR ADVENTURE- BEGIN THEIR VOYAGE...
SO, CHIEF - HOW FAST DOES THIS LITTLE RED BALL GO? DOES SHE HAVE SOME HORSEPOWER?
SHE GOES THE POSTED SPEED LIMIT.
MY AWESOME HAIR IS BARELY BLOWIN' IN THE WIND!
ABOUT THAT HAIR... I'M NOT SURE IT WILL PASS INSPECTION...
AGENTS OF ADVENTURE HAVE STRICT STANDARDS.
WELL NOW- THAT HUGE CHUNK OF CHEDDAR IN THE TRUNK? MAY-BE IT WON'T PASS INSPECTION EITHER...
!
YOU KNOW WHAT? I THINK YOUR HAIR LOOKS JUST FINE!

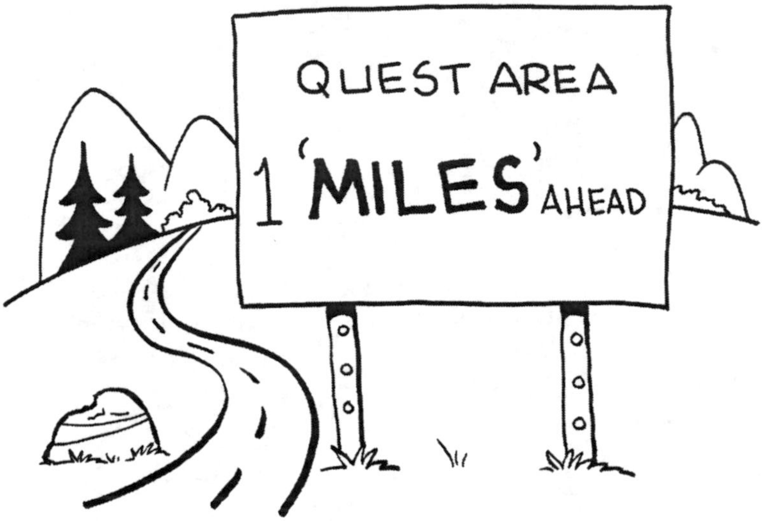
QUEST AREA
1 'MILES' AHEAD

Chapter Three

Inching Forward

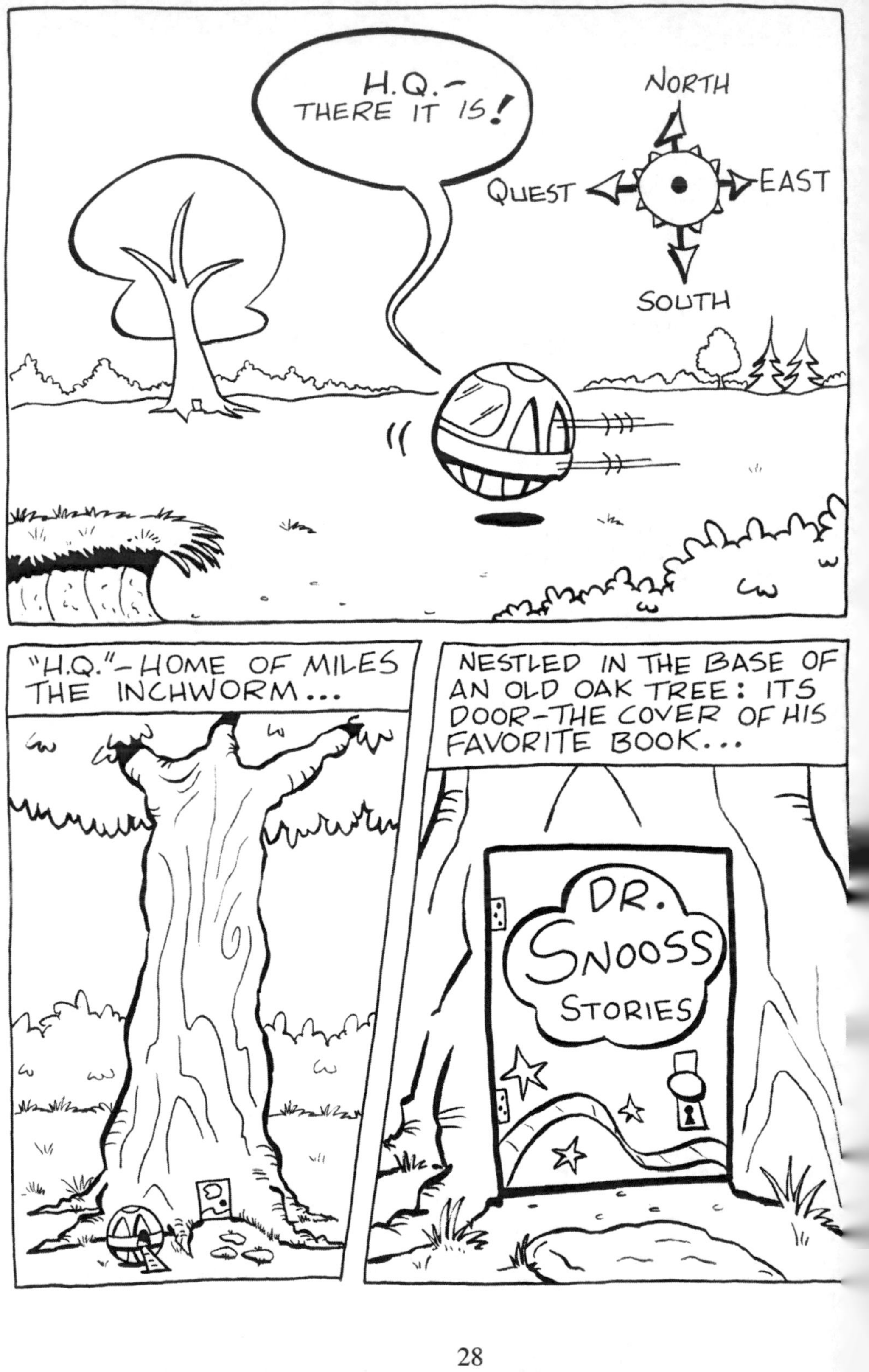
H.Q.–
THERE IT IS!
NORTH
QUEST
EAST
SOUTH
"H.Q."–HOME OF MILES
THE INCHWORM...
NESTLED IN THE BASE OF
AN OLD OAK TREE: ITS
DOOR–THE COVER OF HIS
FAVORITE BOOK...
DR.
SNOOSS
STORIES

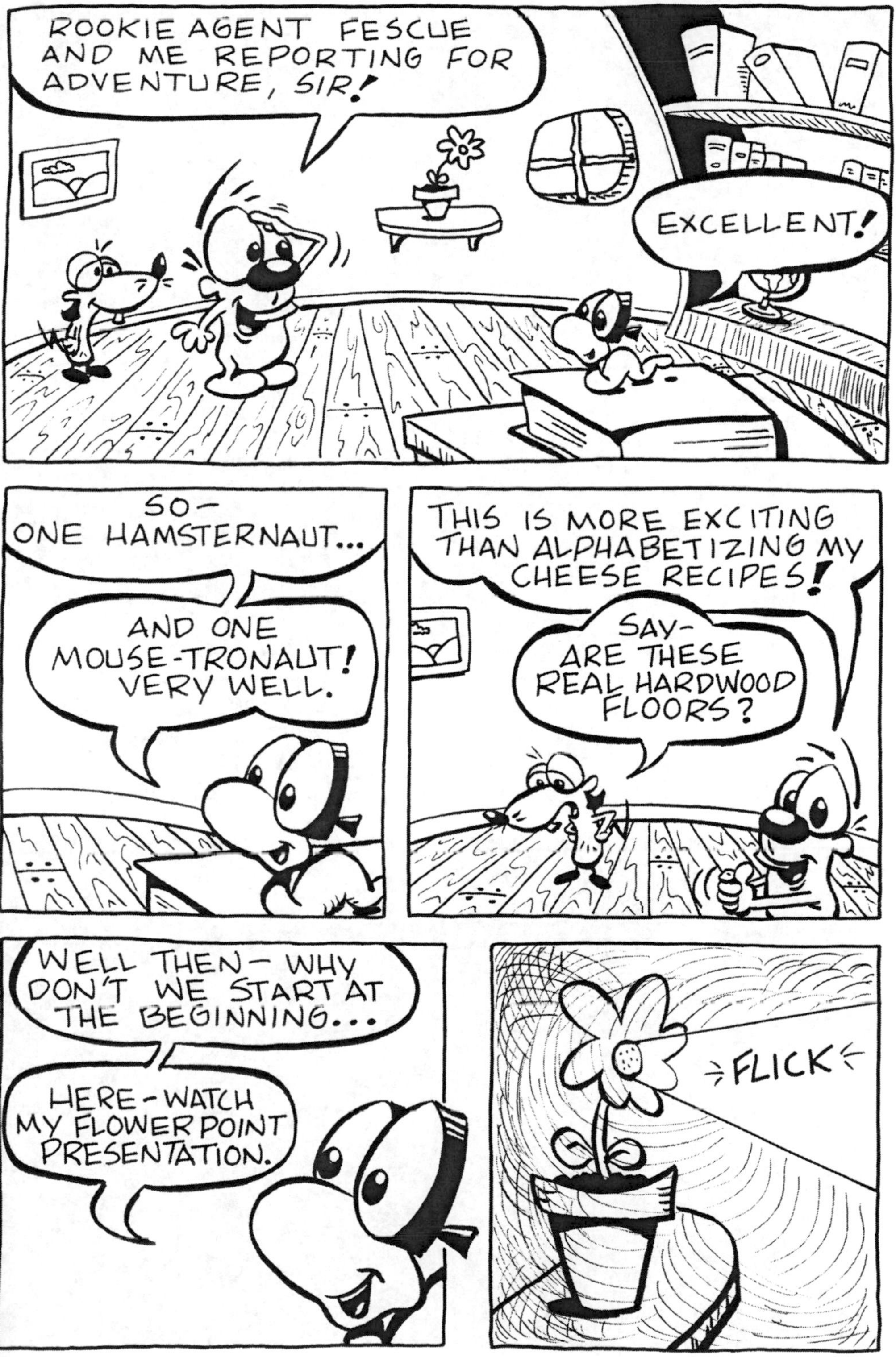
ROOKIE AGENT FESCUE AND ME REPORTING FOR ADVENTURE, SIR!
EXCELLENT!
SO- ONE HAMSTERNAUT...
AND ONE MOUSE-TRONAUT! VERY WELL.
THIS IS MORE EXCITING THAN ALPHABETIZING MY CHEESE RECIPES!
SAY- ARE THESE REAL HARDWOOD FLOORS?
WELL THEN- WHY DON'T WE START AT THE BEGINNING...
HERE- WATCH MY FLOWER POINT PRESENTATION.
FLICK

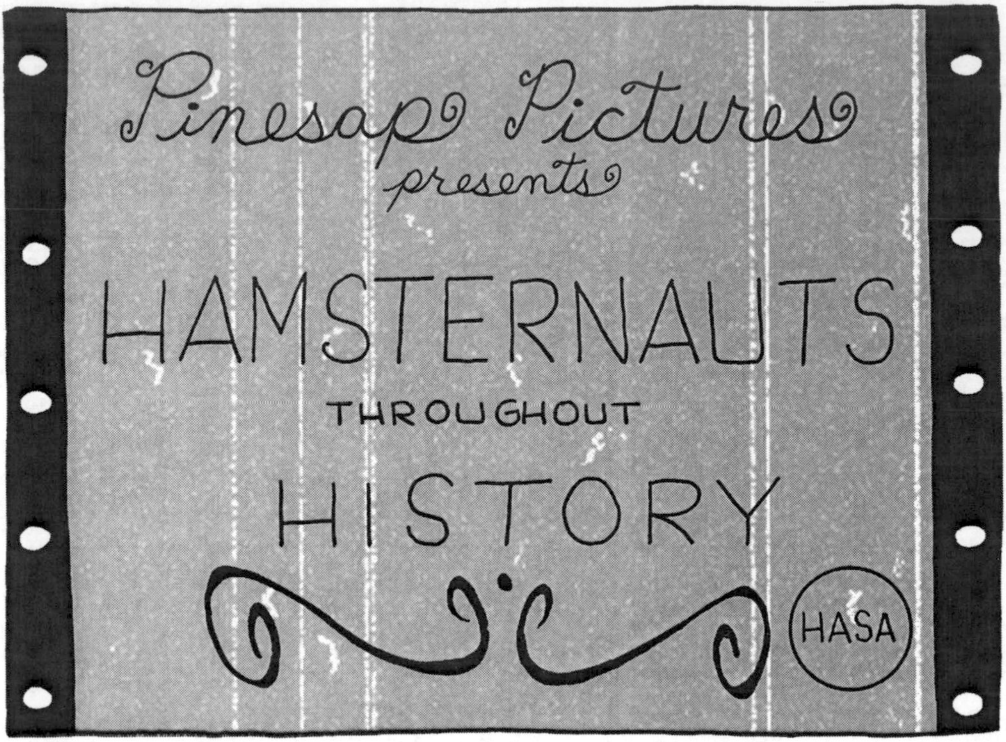
Pinesap Pictures
presents
HAMSTERNAUTS
THROUGHOUT
HISTORY
HASA

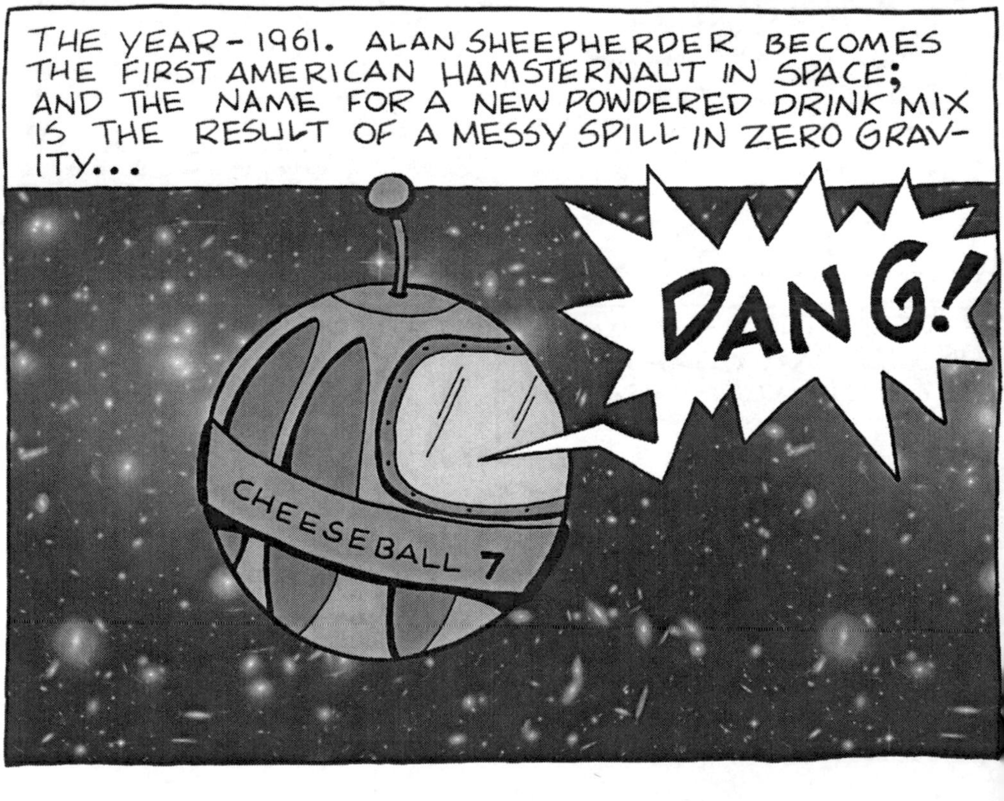
THE YEAR - 1961. ALAN SHEEPHERDER BECOMES THE FIRST AMERICAN HAMSTERNAUT IN SPACE; AND THE NAME FOR A NEW POWDERED DRINK MIX IS THE RESULT OF A MESSY SPILL IN ZERO GRAVITY...
DANG!
CHEESEBALL 7

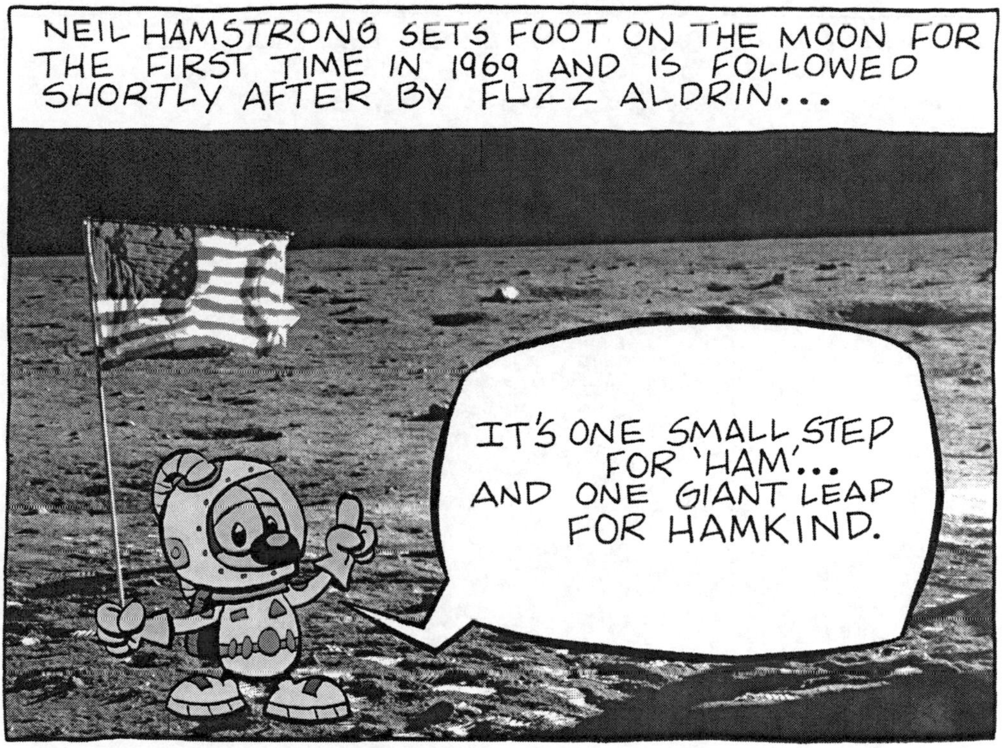
NEIL HAMSTRONG SETS FOOT ON THE MOON FOR THE FIRST TIME IN 1969 AND IS FOLLOWED SHORTLY AFTER BY FUZZ ALDRIN...
IT'S ONE SMALL STEP FOR 'HAM'... AND ONE GIANT LEAP FOR HAMKIND.

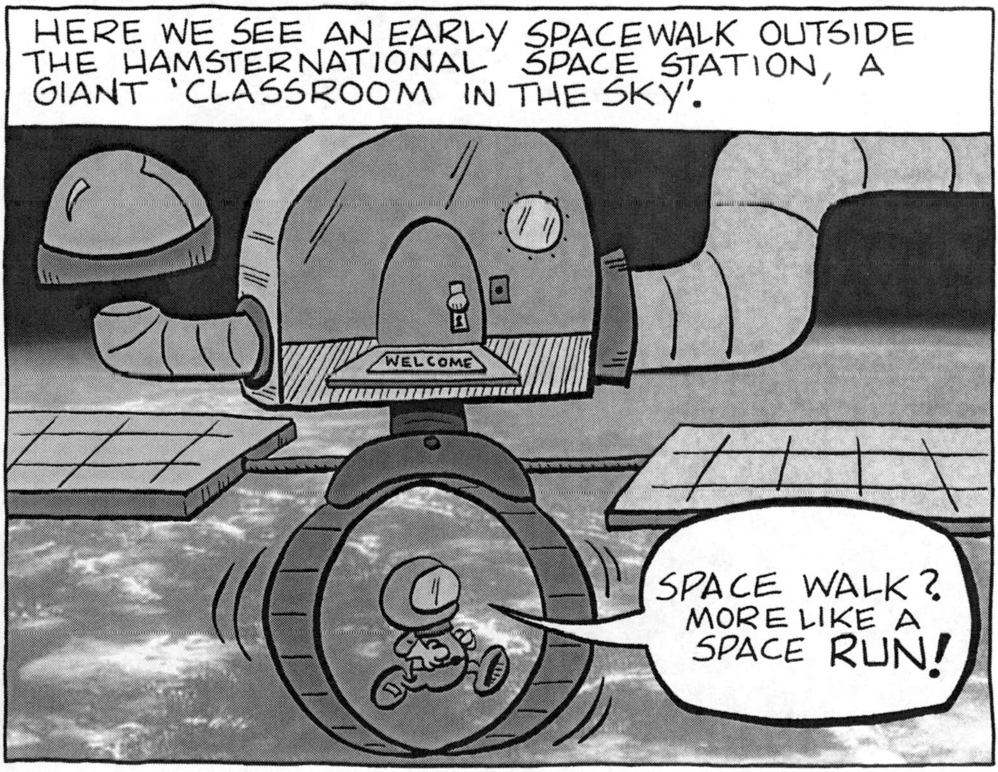
HERE WE SEE AN EARLY SPACEWALK OUTSIDE THE HAMSTERNATIONAL SPACE STATION, A GIANT 'CLASSROOM IN THE SKY'.
WELCOME
SPACE WALK? MORE LIKE A SPACE RUN!

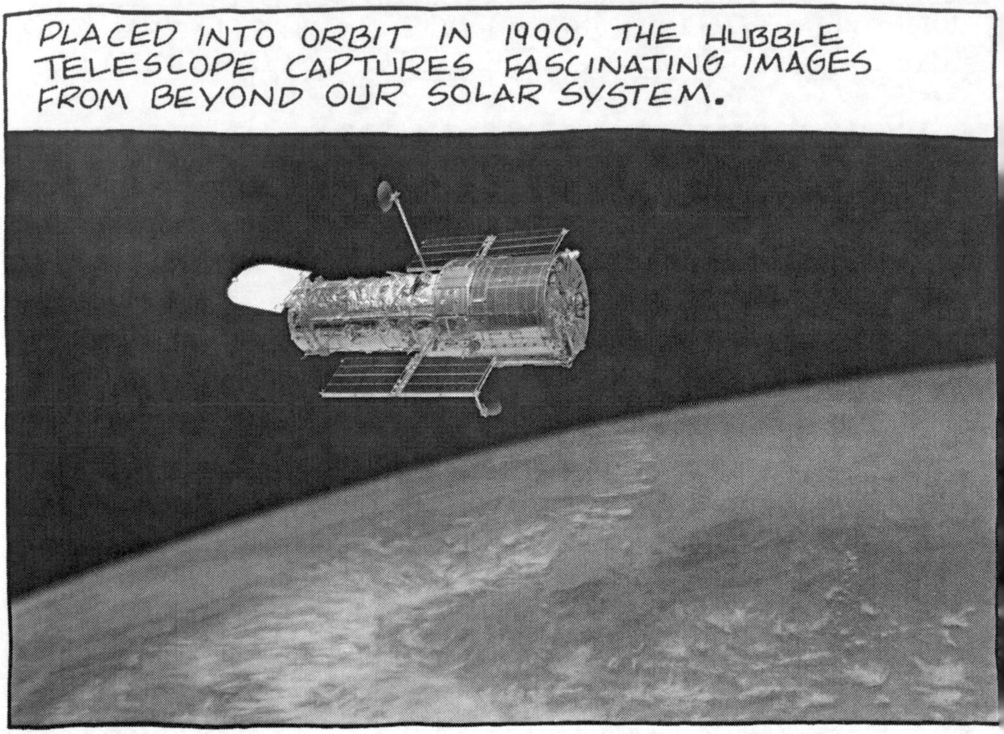
PLACED INTO ORBIT IN 1990, THE HUBBLE TELESCOPE CAPTURES FASCINATING IMAGES FROM BEYOND OUR SOLAR SYSTEM.

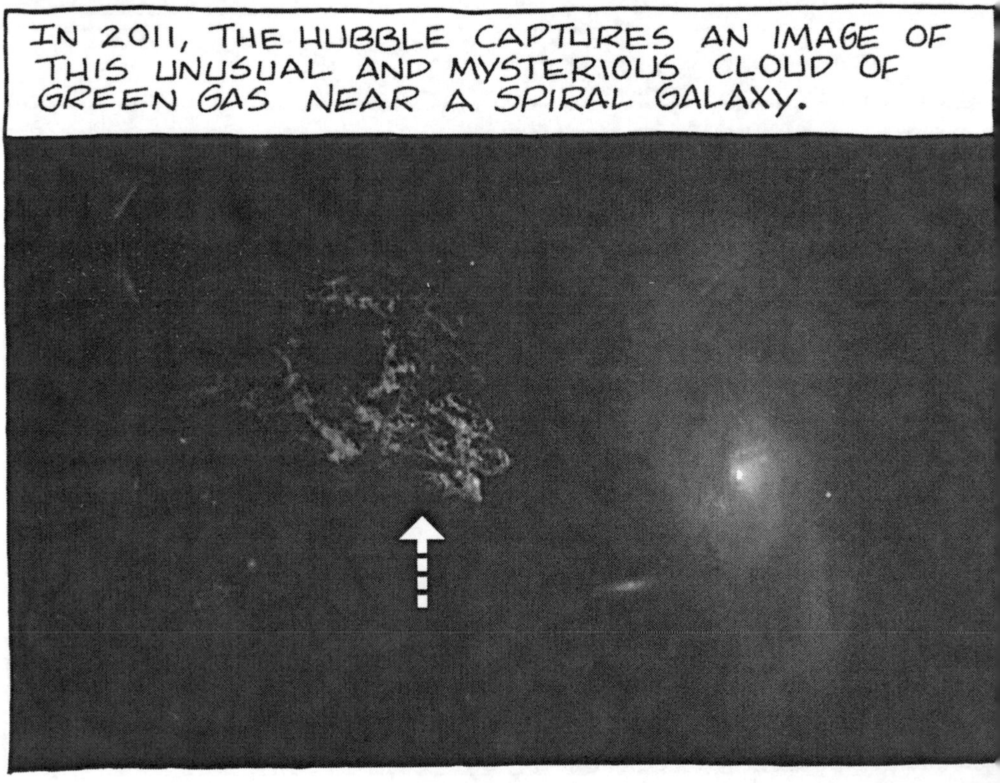
IN 2011, THE HUBBLE CAPTURES AN IMAGE OF THIS UNUSUAL AND MYSTERIOUS CLOUD OF GREEN GAS NEAR A SPIRAL GALAXY.

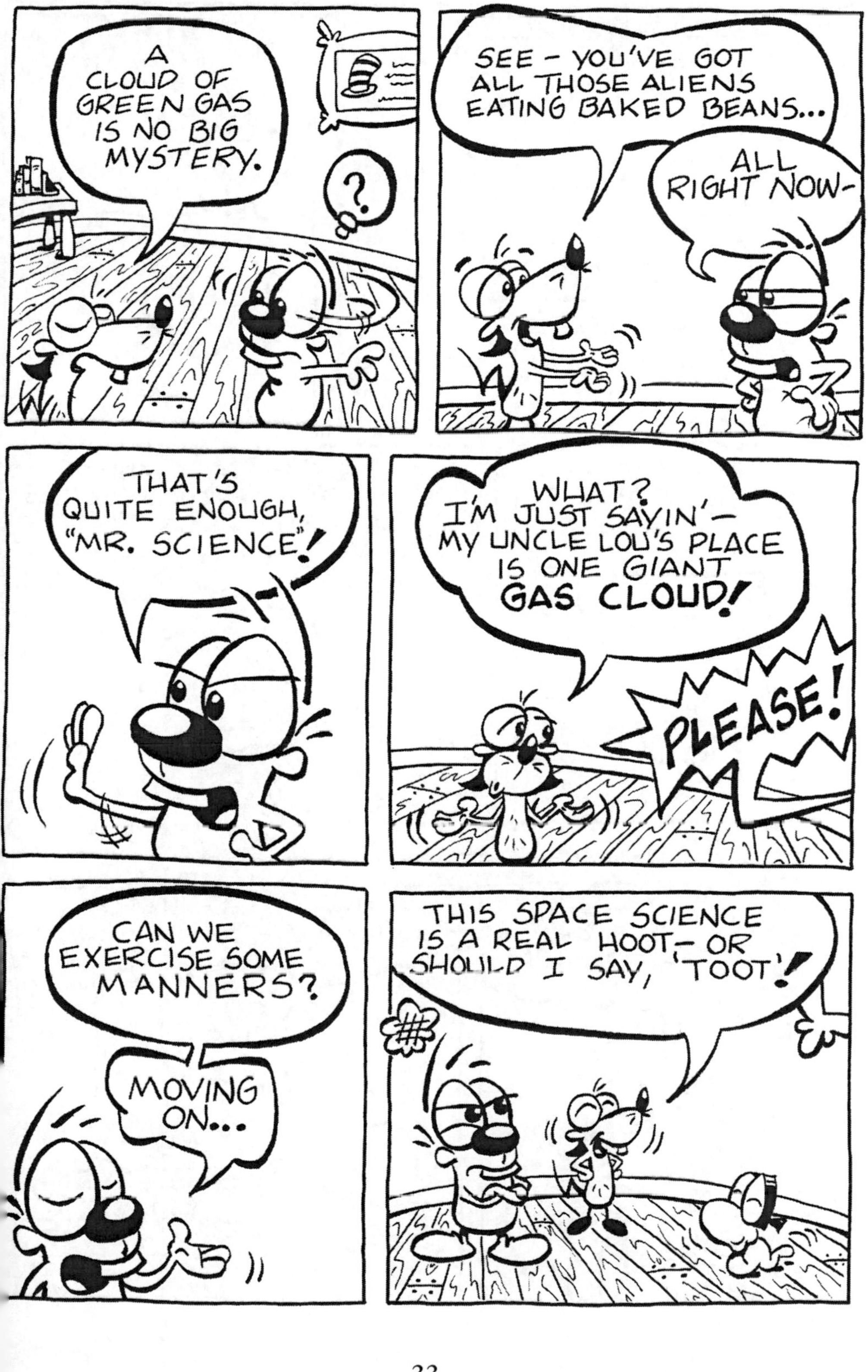
A CLOUD OF GREEN GAS IS NO BIG MYSTERY.
?
SEE - YOU'VE GOT ALL THOSE ALIENS EATING BAKED BEANS...
ALL RIGHT NOW-
THAT'S QUITE ENOUGH, "MR. SCIENCE"!
WHAT? I'M JUST SAYIN'— MY UNCLE LOU'S PLACE IS ONE GIANT GAS CLOUD!
PLEASE!
CAN WE EXERCISE SOME MANNERS?
MOVING ON...
THIS SPACE SCIENCE IS A REAL HOOT- OR SHOULD I SAY, 'TOOT'!

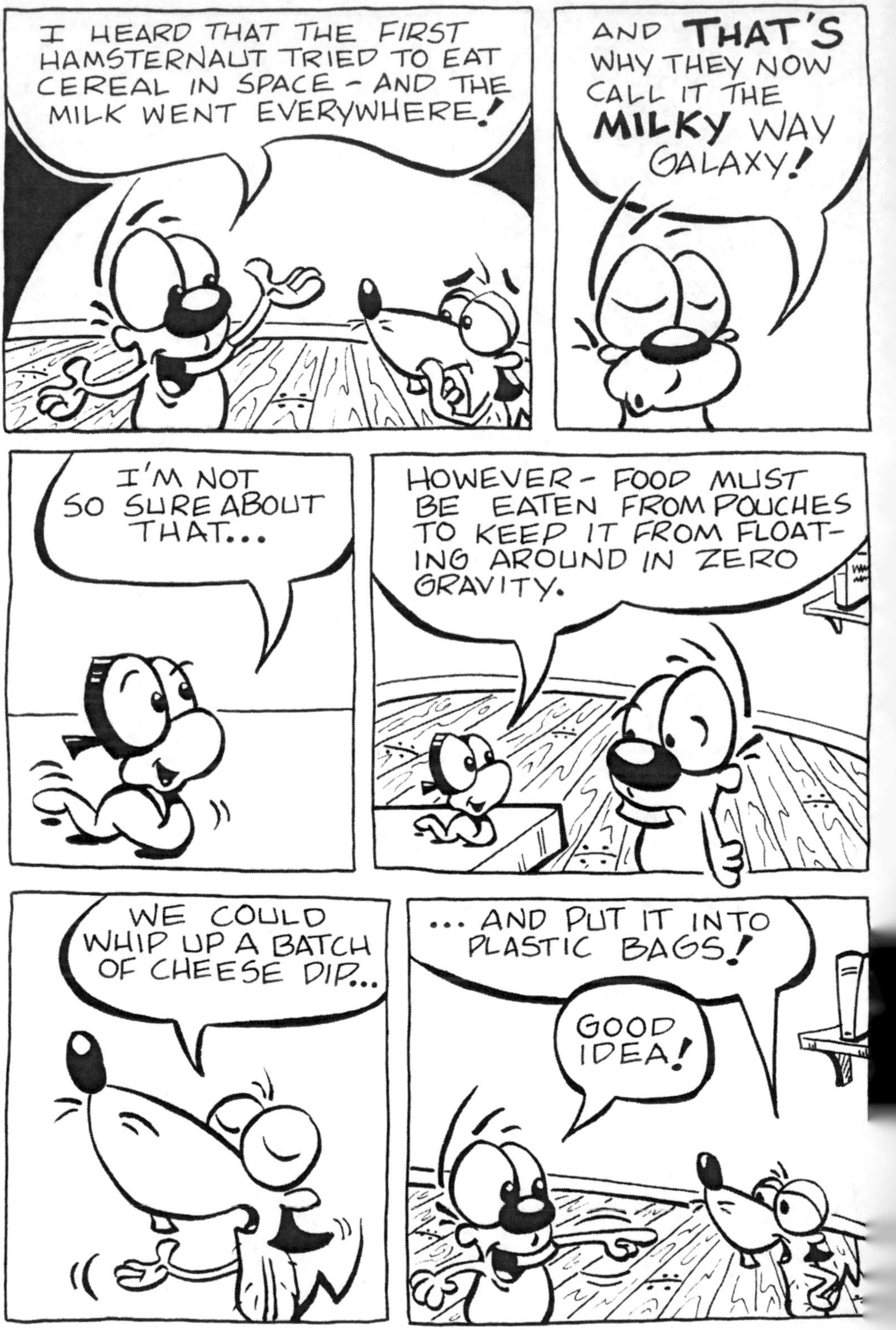
I HEARD THAT THE FIRST HAMSTERNAUT TRIED TO EAT CEREAL IN SPACE - AND THE MILK WENT EVERYWHERE!
AND THAT'S WHY THEY NOW CALL IT THE MILKY WAY GALAXY!
I'M NOT SO SURE ABOUT THAT...
HOWEVER- FOOD MUST BE EATEN FROM POUCHES TO KEEP IT FROM FLOATING AROUND IN ZERO GRAVITY.
WE COULD WHIP UP A BATCH OF CHEESE DIP...
... AND PUT IT INTO PLASTIC BAGS!
GOOD IDEA!

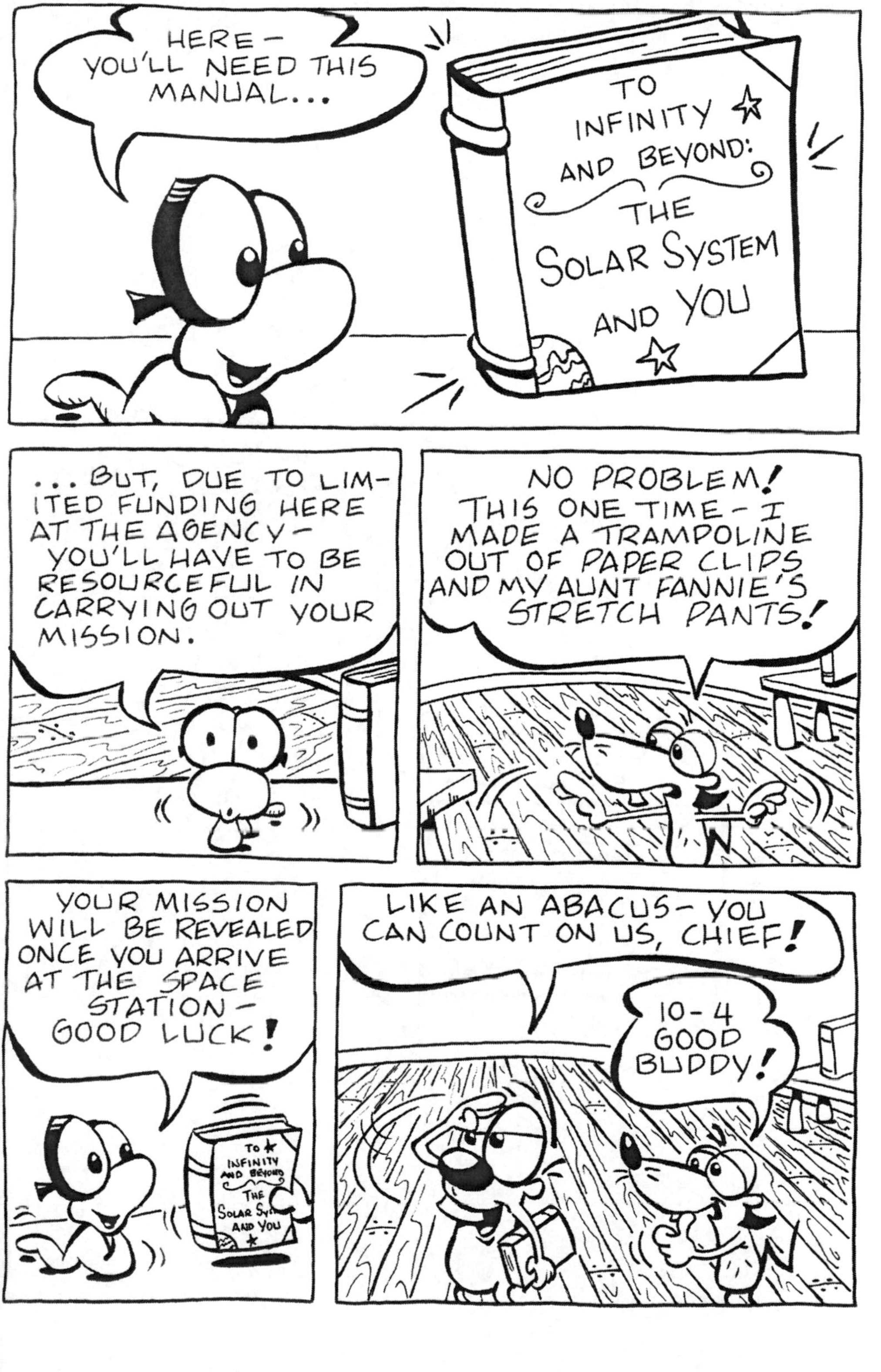
HERE - YOU'LL NEED THIS MANUAL...
TO INFINITY AND BEYOND: THE SOLAR SYSTEM AND YOU
...BUT, DUE TO LIMITED FUNDING HERE AT THE AGENCY - YOU'LL HAVE TO BE RESOURCEFUL IN CARRYING OUT YOUR MISSION.
NO PROBLEM! THIS ONE TIME - I MADE A TRAMPOLINE OUT OF PAPER CLIPS AND MY AUNT FANNIE'S STRETCH PANTS!
YOUR MISSION WILL BE REVEALED ONCE YOU ARRIVE AT THE SPACE STATION - GOOD LUCK!
TO INFINITY AND BEYOND THE SOLAR SY AND YOU
LIKE AN ABACUS - YOU CAN COUNT ON US, CHIEF!
10-4 GOOD BUDDY!

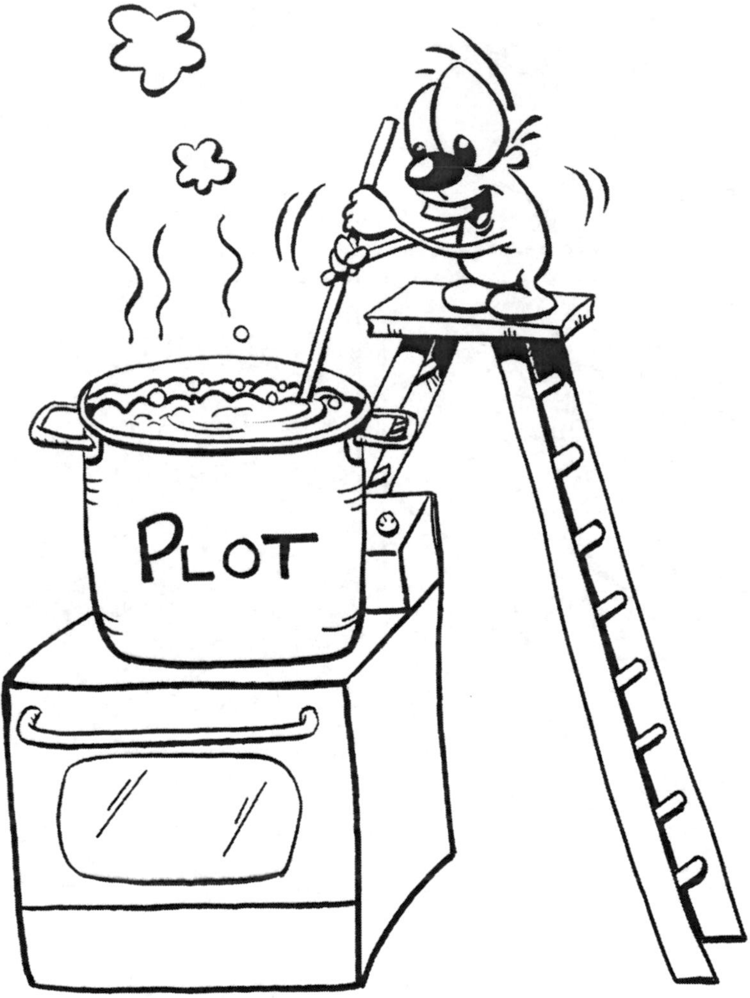
PLOT

Chapter Four

And The Plot Thickens

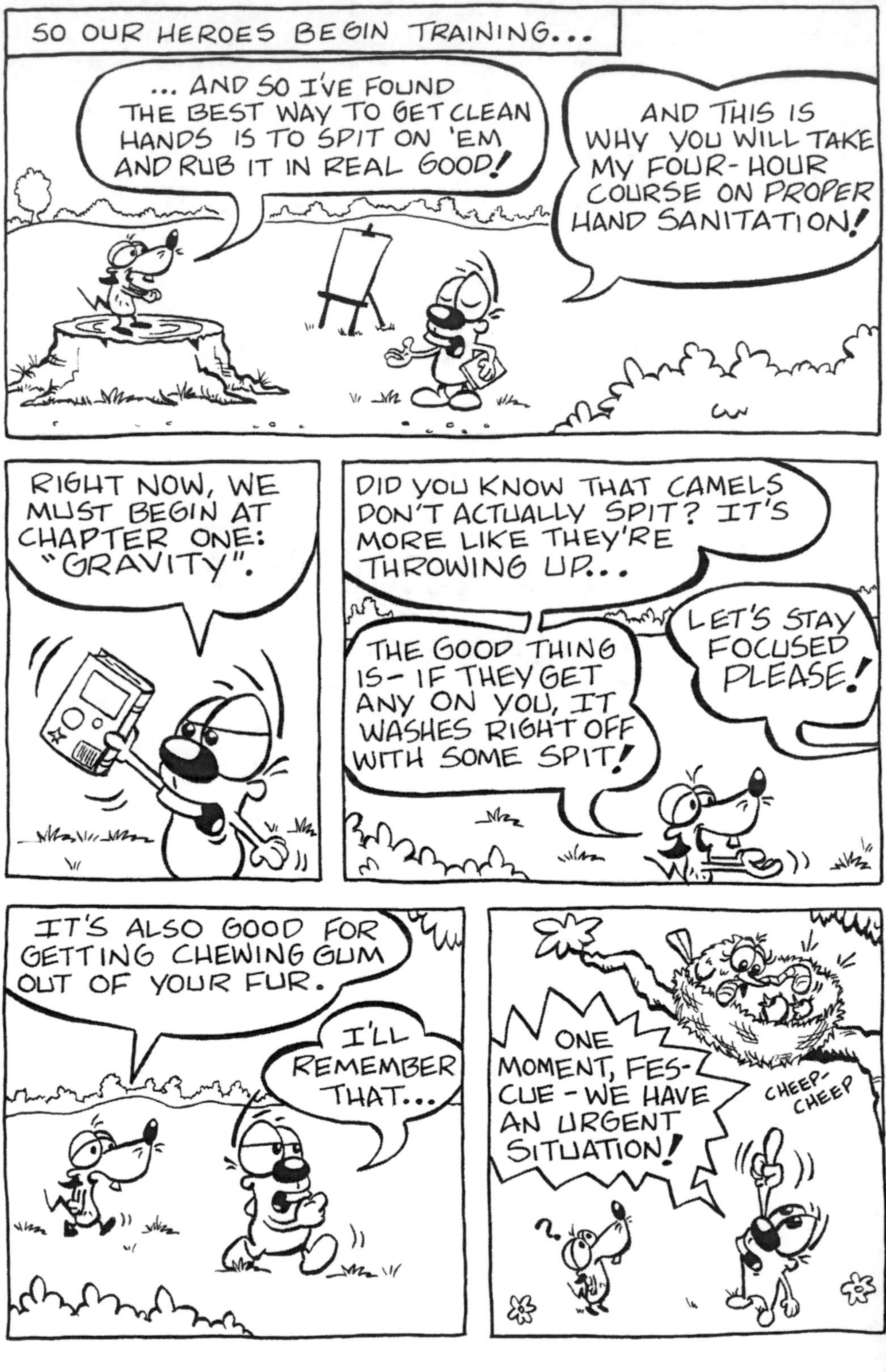
SO OUR HEROES BEGIN TRAINING...
... AND SO I'VE FOUND THE BEST WAY TO GET CLEAN HANDS IS TO SPIT ON 'EM AND RUB IT IN REAL GOOD!
AND THIS IS WHY YOU WILL TAKE MY FOUR-HOUR COURSE ON PROPER HAND SANITATION!
RIGHT NOW, WE MUST BEGIN AT CHAPTER ONE: "GRAVITY".
DID YOU KNOW THAT CAMELS DON'T ACTUALLY SPIT? IT'S MORE LIKE THEY'RE THROWING UP...
THE GOOD THING IS- IF THEY GET ANY ON YOU, IT WASHES RIGHT OFF WITH SOME SPIT!
LET'S STAY FOCUSED PLEASE!
IT'S ALSO GOOD FOR GETTING CHEWING GUM OUT OF YOUR FUR.
I'LL REMEMBER THAT...
ONE MOMENT, FES-CUE - WE HAVE AN URGENT SITUATION!
CHEEP-CHEEP

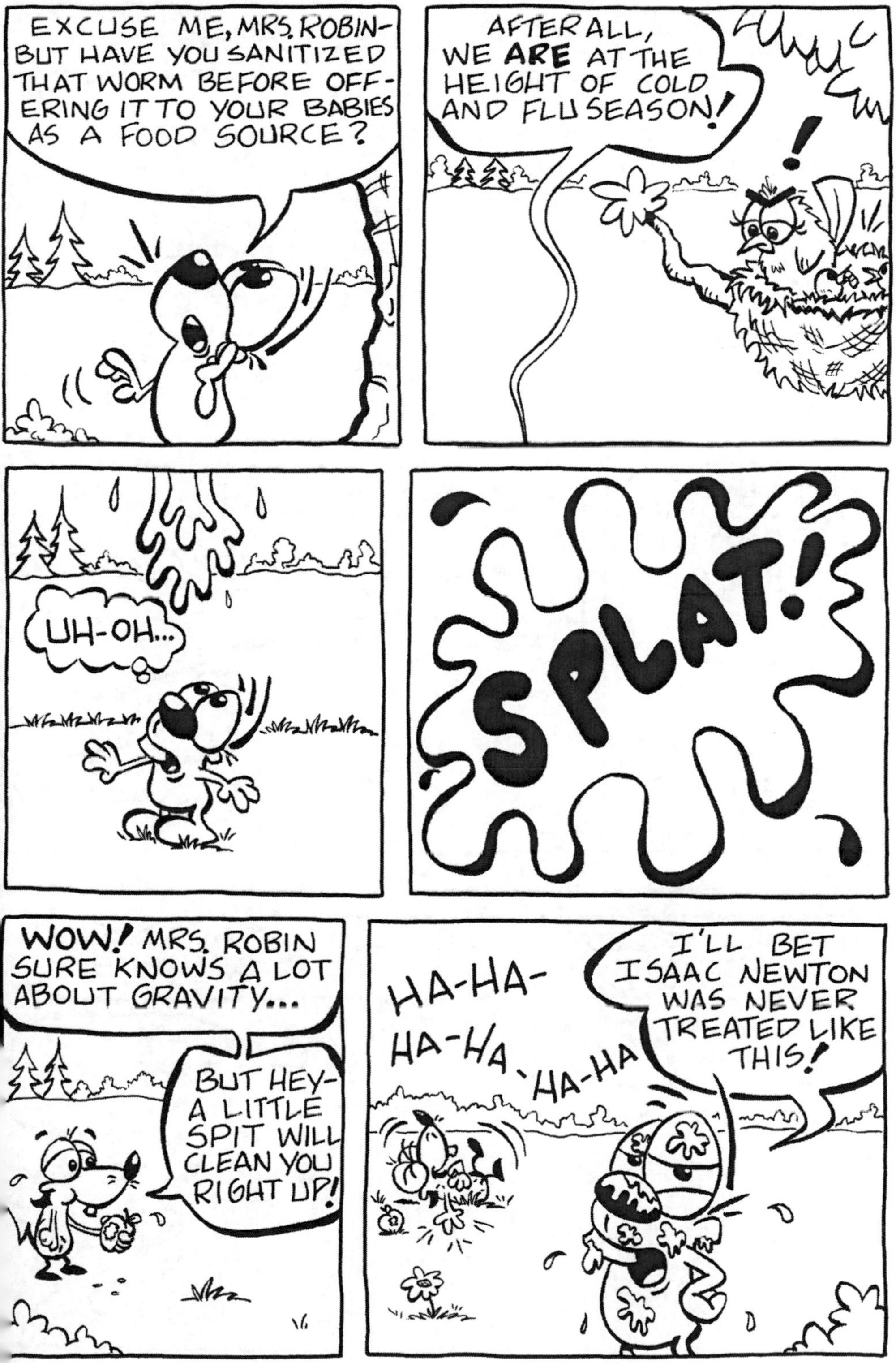
EXCUSE ME, MRS. ROBIN- BUT HAVE YOU SANITIZED THAT WORM BEFORE OFF-ERING IT TO YOUR BABIES AS A FOOD SOURCE?
AFTER ALL, WE **ARE** AT THE HEIGHT OF COLD AND FLU SEASON!
!
UH-OH...
SPLAT!
WOW! MRS. ROBIN SURE KNOWS A LOT ABOUT GRAVITY...
BUT HEY- A LITTLE SPIT WILL CLEAN YOU RIGHT UP!
HA-HA- HA-HA-HA-HA
I'LL BET ISAAC NEWTON WAS NEVER TREATED LIKE THIS!

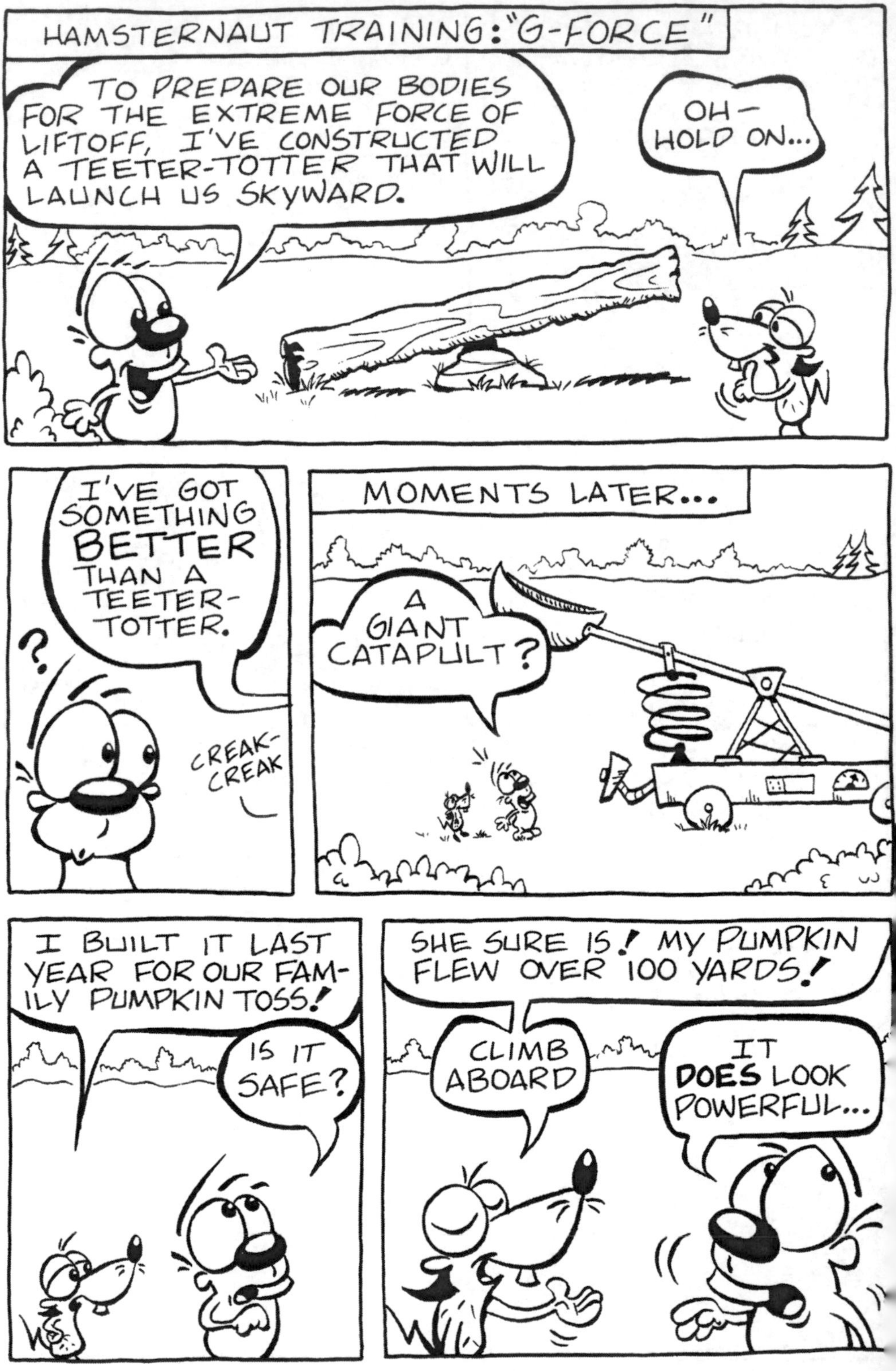
HAMSTERNAUT TRAINING: "G-FORCE"
TO PREPARE OUR BODIES FOR THE EXTREME FORCE OF LIFTOFF, I'VE CONSTRUCTED A TEETER-TOTTER THAT WILL LAUNCH US SKYWARD.
OH— HOLD ON...
I'VE GOT SOMETHING BETTER THAN A TEETER-TOTTER.
?
CREAK-CREAK
MOMENTS LATER...
A GIANT CATAPULT?
I BUILT IT LAST YEAR FOR OUR FAMILY PUMPKIN TOSS!
IS IT SAFE?
SHE SURE IS! MY PUMPKIN FLEW OVER 100 YARDS!
CLIMB ABOARD
IT DOES LOOK POWERFUL...

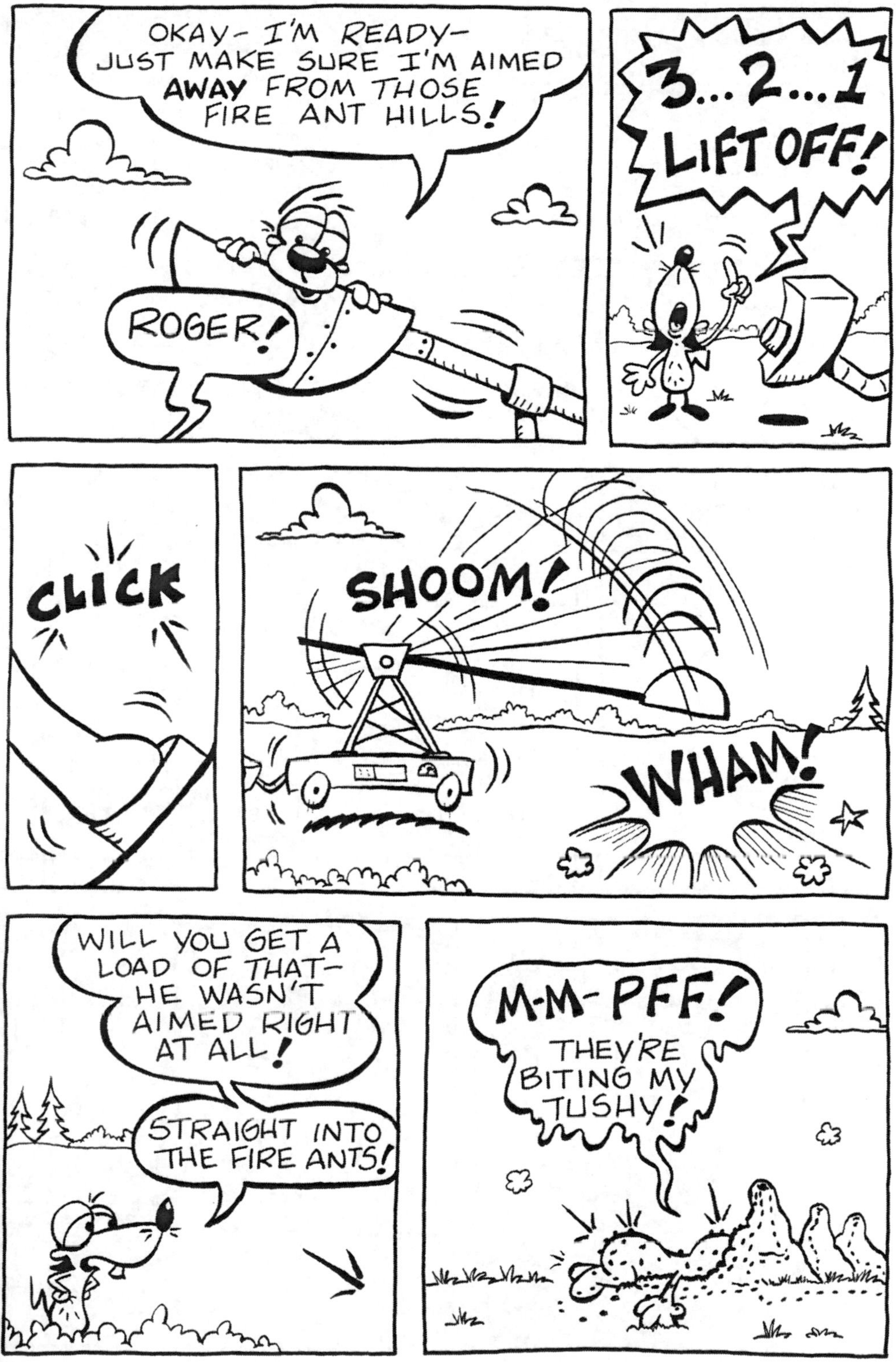
OKAY- I'M READY- JUST MAKE SURE I'M AIMED **AWAY** FROM THOSE FIRE ANT HILLS!
ROGER!
3... 2... 1 LIFT OFF!
CLICK
SHOOM!
WHAM!
WILL YOU GET A LOAD OF THAT- HE WASN'T AIMED RIGHT AT ALL!
STRAIGHT INTO THE FIRE ANTS!
M-M-PFF! THEY'RE BITING MY TUSHY!

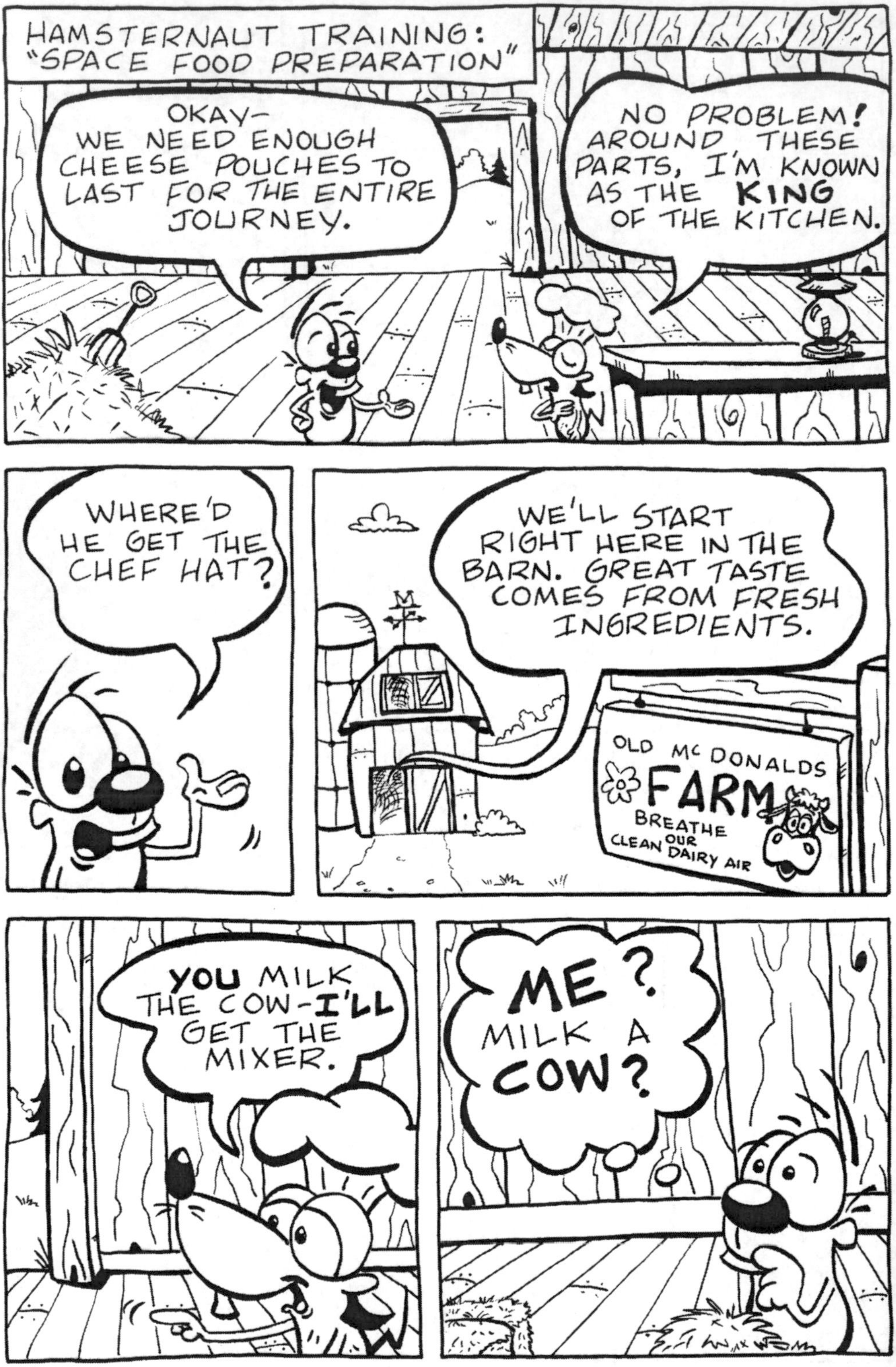
HAMSTERNAUT TRAINING: "SPACE FOOD PREPARATION"
OKAY- WE NEED ENOUGH CHEESE POUCHES TO LAST FOR THE ENTIRE JOURNEY.
NO PROBLEM! AROUND THESE PARTS, I'M KNOWN AS THE KING OF THE KITCHEN.
WHERE'D HE GET THE CHEF HAT?
WE'LL START RIGHT HERE IN THE BARN. GREAT TASTE COMES FROM FRESH INGREDIENTS.
OLD McDONALDS FARM
BREATHE OUR CLEAN DAIRY AIR
YOU MILK THE COW-I'LL GET THE MIXER.
ME? MILK A COW?

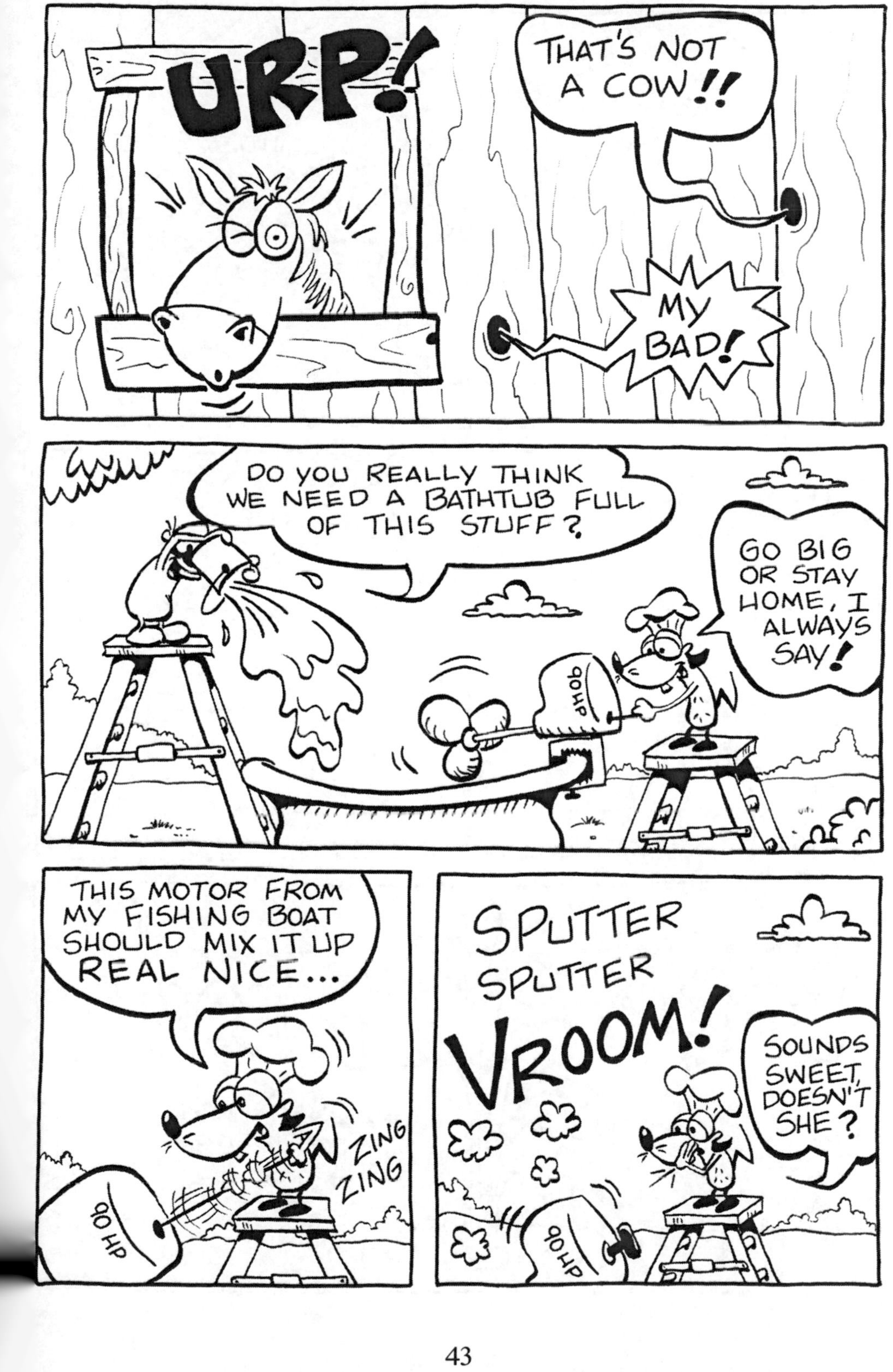
URP!
THAT'S NOT A COW!!
MY BAD!
DO YOU REALLY THINK WE NEED A BATHTUB FULL OF THIS STUFF?
90HP
GO BIG OR STAY HOME, I ALWAYS SAY!
THIS MOTOR FROM MY FISHING BOAT SHOULD MIX IT UP REAL NICE...
ZING ZING
90HP
SPUTTER
SPUTTER
VROOM!
SOUNDS SWEET, DOESN'T SHE?
90HP

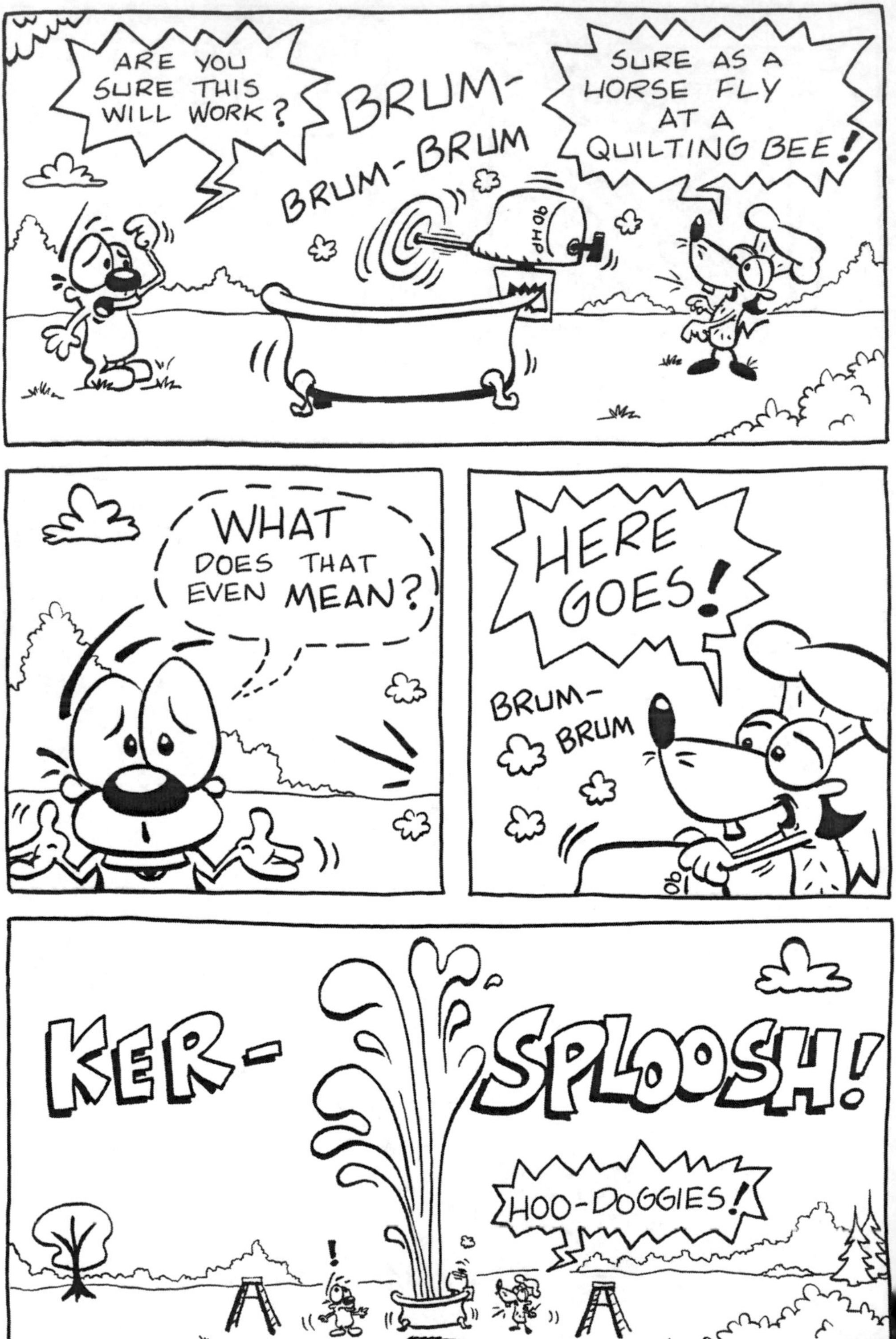
ARE YOU SURE THIS WILL WORK?
BRUM-BRUM-BRUM
SURE AS A HORSE FLY AT A QUILTING BEE!
90HP
WHAT DOES THAT EVEN MEAN?
HERE GOES!
BRUM-BRUM
KER-SPLOOSH!
HOO-DOGGIES!

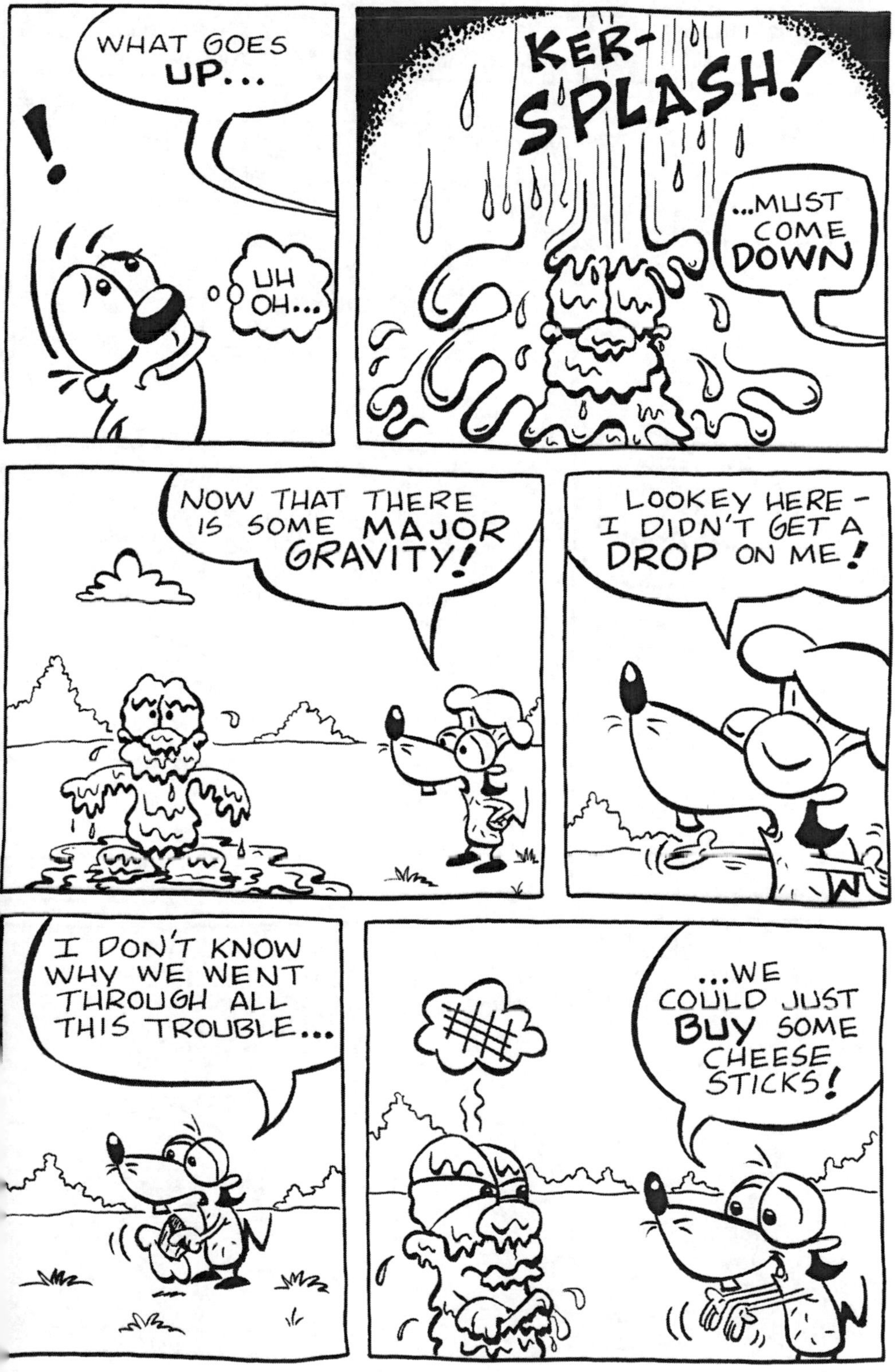
WHAT GOES UP...
!
UH OH...
KER-SPLASH!
...MUST COME DOWN
NOW THAT THERE IS SOME MAJOR GRAVITY!
LOOKEY HERE - I DIDN'T GET A DROP ON ME!
I DON'T KNOW WHY WE WENT THROUGH ALL THIS TROUBLE...
...WE COULD JUST BUY SOME CHEESE STICKS!

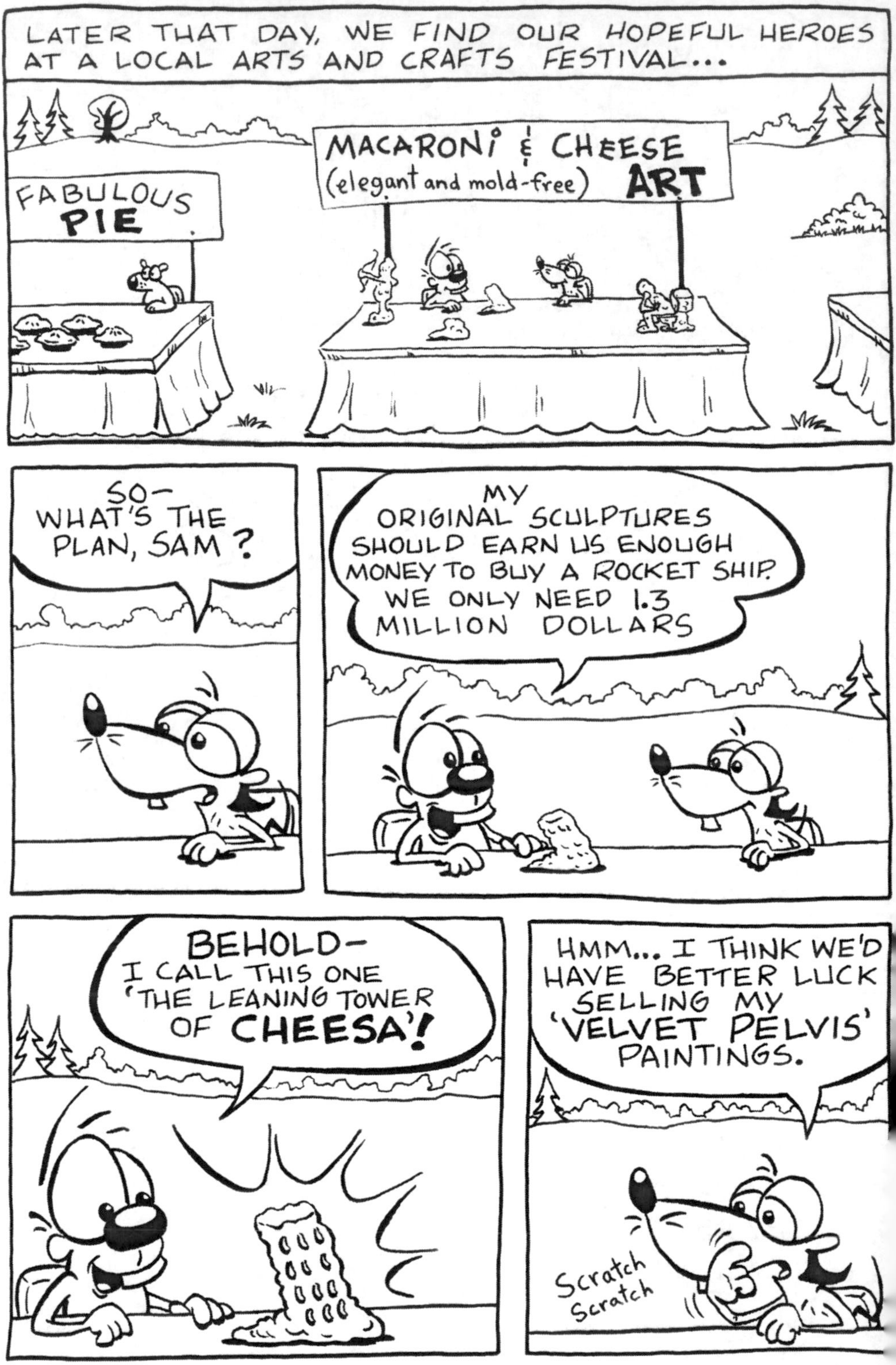
LATER THAT DAY, WE FIND OUR HOPEFUL HEROES AT A LOCAL ARTS AND CRAFTS FESTIVAL...
FABULOUS PIE
MACARONI & CHEESE (elegant and mold-free) ART
SO- WHAT'S THE PLAN, SAM?
MY ORIGINAL SCULPTURES SHOULD EARN US ENOUGH MONEY TO BUY A ROCKET SHIP. WE ONLY NEED 1.3 MILLION DOLLARS
BEHOLD- I CALL THIS ONE 'THE LEANING TOWER OF CHEESA'!
HMM... I THINK WE'D HAVE BETTER LUCK SELLING MY 'VELVET PELVIS' PAINTINGS.
Scratch Scratch

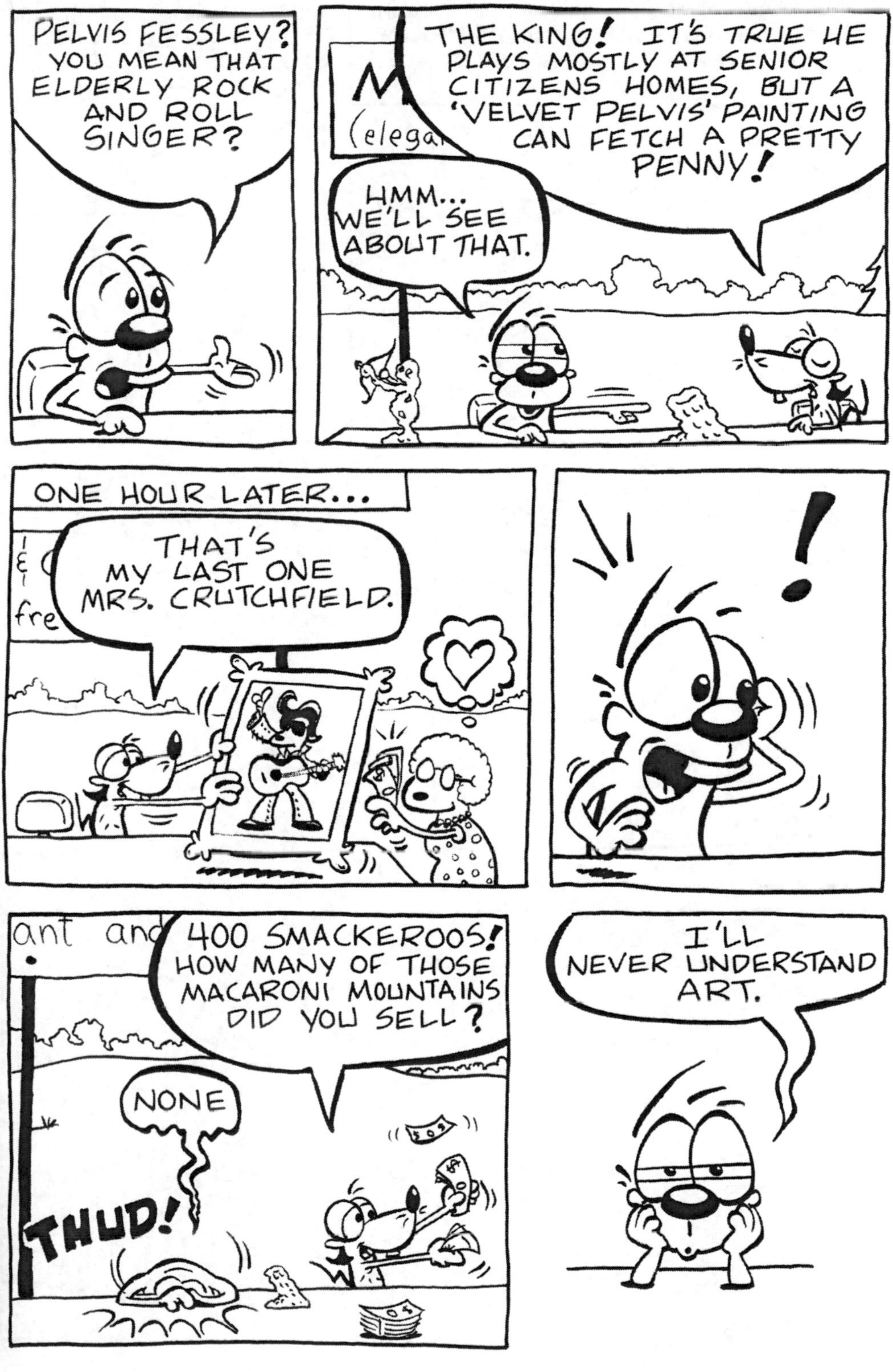
PELVIS FESSLEY? YOU MEAN THAT ELDERLY ROCK AND ROLL SINGER?
THE KING! IT'S TRUE HE PLAYS MOSTLY AT SENIOR CITIZENS HOMES, BUT A 'VELVET PELVIS' PAINTING CAN FETCH A PRETTY PENNY!
HMM... WE'LL SEE ABOUT THAT.
ONE HOUR LATER...
THAT'S MY LAST ONE MRS. CRUTCHFIELD.
!
400 SMACKEROOS! HOW MANY OF THOSE MACARONI MOUNTAINS DID YOU SELL?
NONE
THUD!
I'LL NEVER UNDERSTAND ART.

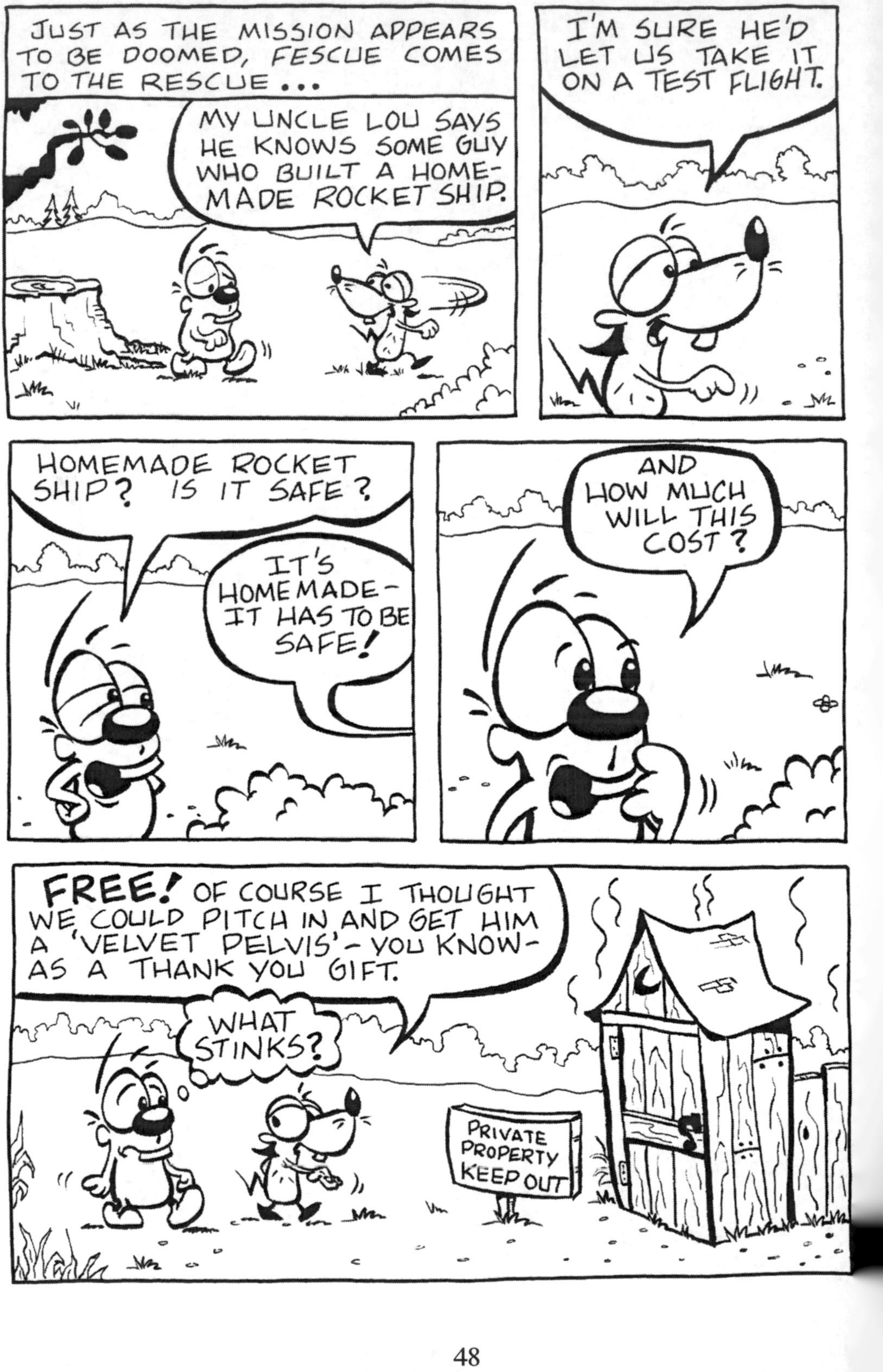
JUST AS THE MISSION APPEARS TO BE DOOMED, FESCUE COMES TO THE RESCUE...
MY UNCLE LOU SAYS HE KNOWS SOME GUY WHO BUILT A HOME-MADE ROCKET SHIP.
I'M SURE HE'D LET US TAKE IT ON A TEST FLIGHT.
HOMEMADE ROCKET SHIP? IS IT SAFE?
IT'S HOMEMADE- IT HAS TO BE SAFE!
AND HOW MUCH WILL THIS COST?
FREE! OF COURSE I THOUGHT WE COULD PITCH IN AND GET HIM A 'VELVET PELVIS'- YOU KNOW- AS A THANK YOU GIFT.
WHAT STINKS?
PRIVATE PROPERTY KEEP OUT

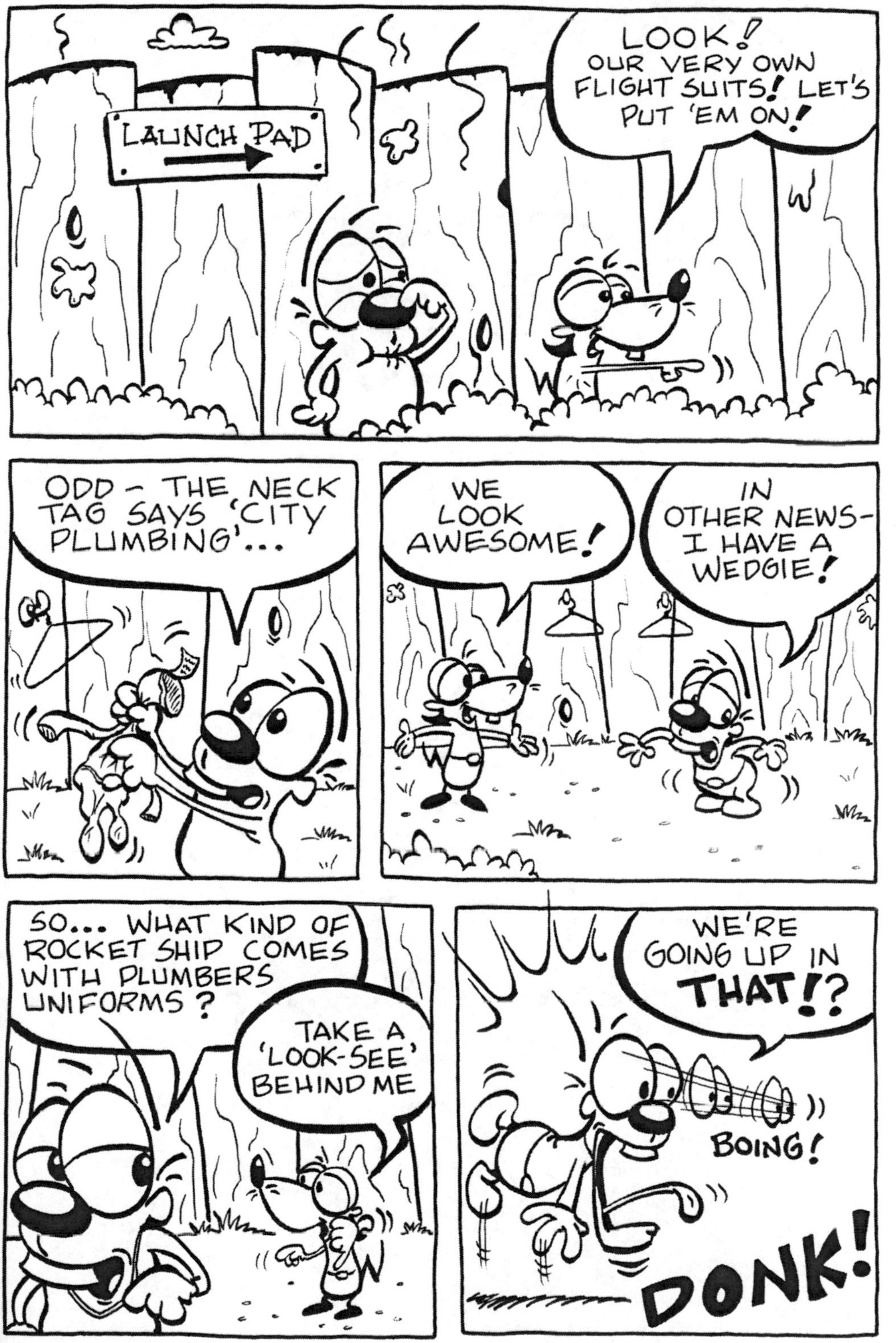

LAUNCH PAD
LOOK! OUR VERY OWN FLIGHT SUITS! LET'S PUT 'EM ON!
ODD - THE NECK TAG SAYS 'CITY PLUMBING'...
WE LOOK AWESOME!
IN OTHER NEWS- I HAVE A WEDGIE!
SO... WHAT KIND OF ROCKET SHIP COMES WITH PLUMBERS UNIFORMS?
TAKE A 'LOOK-SEE' BEHIND ME
WE'RE GOING UP IN THAT!?
BOING!
DONK!

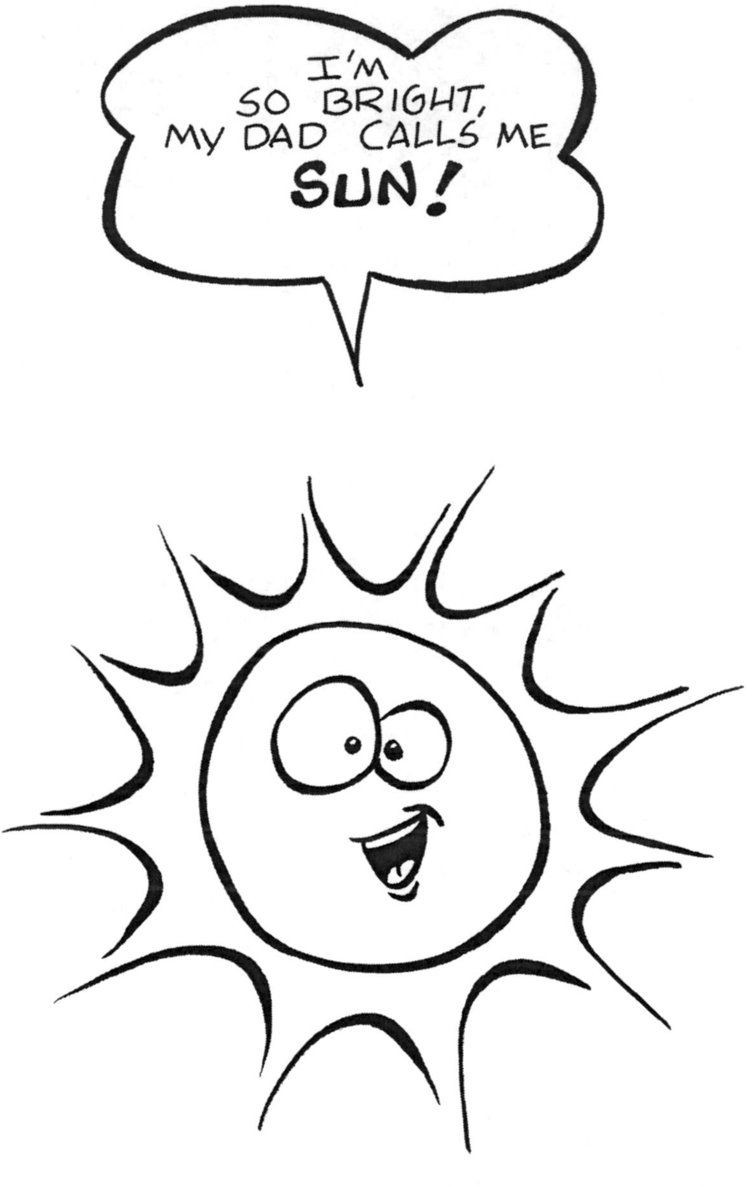
I'M
SO BRIGHT,
MY DAD CALLS ME
SUN!

Chapter Five

Up, Up and
Oy Vey!

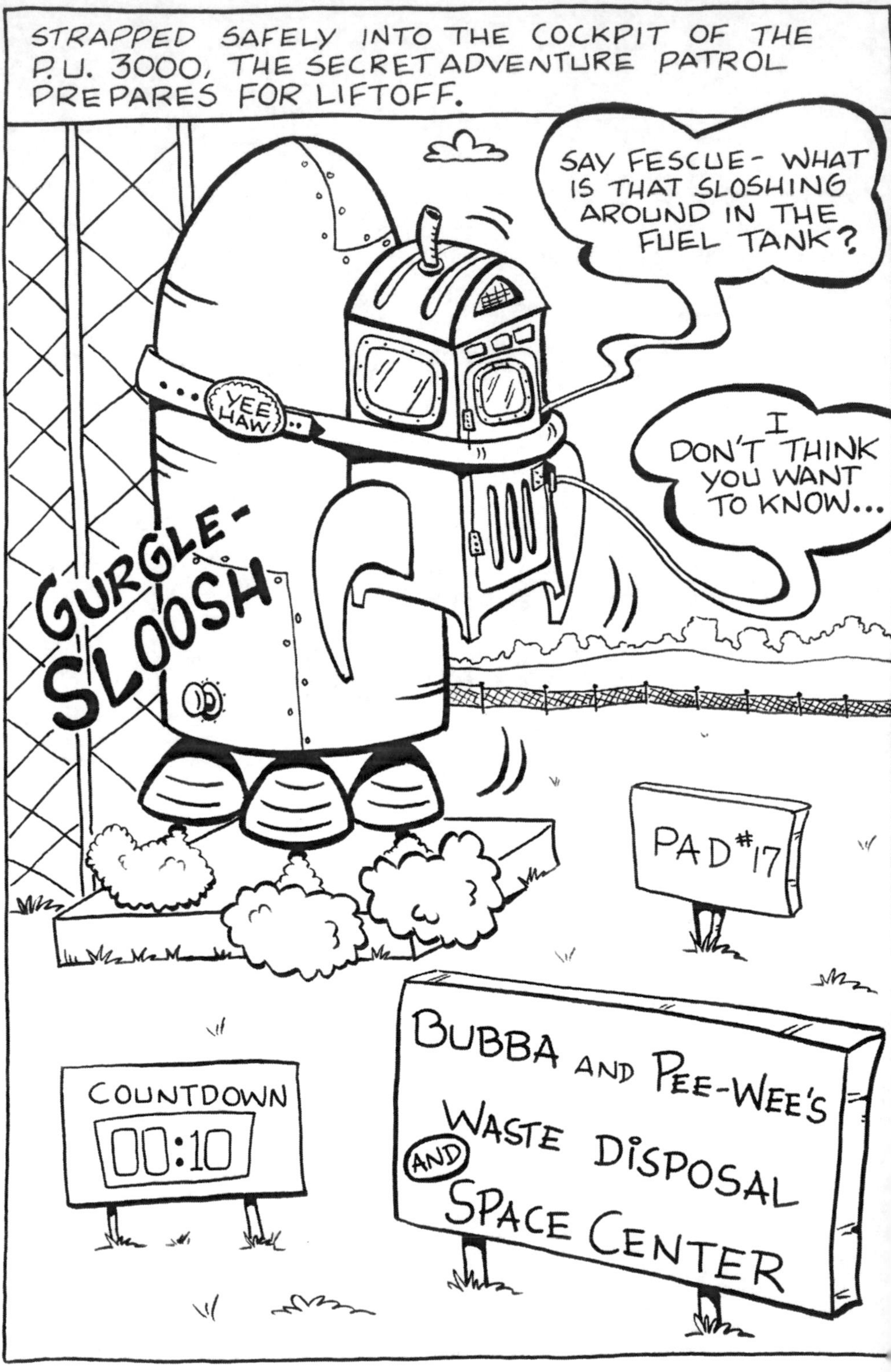
STRAPPED SAFELY INTO THE COCKPIT OF THE P.U. 3000, THE SECRET ADVENTURE PATROL PREPARES FOR LIFTOFF.
SAY FESCUE- WHAT IS THAT SLOSHING AROUND IN THE FUEL TANK?
I DON'T THINK YOU WANT TO KNOW...
YEE HAW
GURGLE- SLOOSH
PAD #17
COUNTDOWN
00:10
BUBBA AND PEE-WEE'S WASTE DISPOSAL AND SPACE CENTER

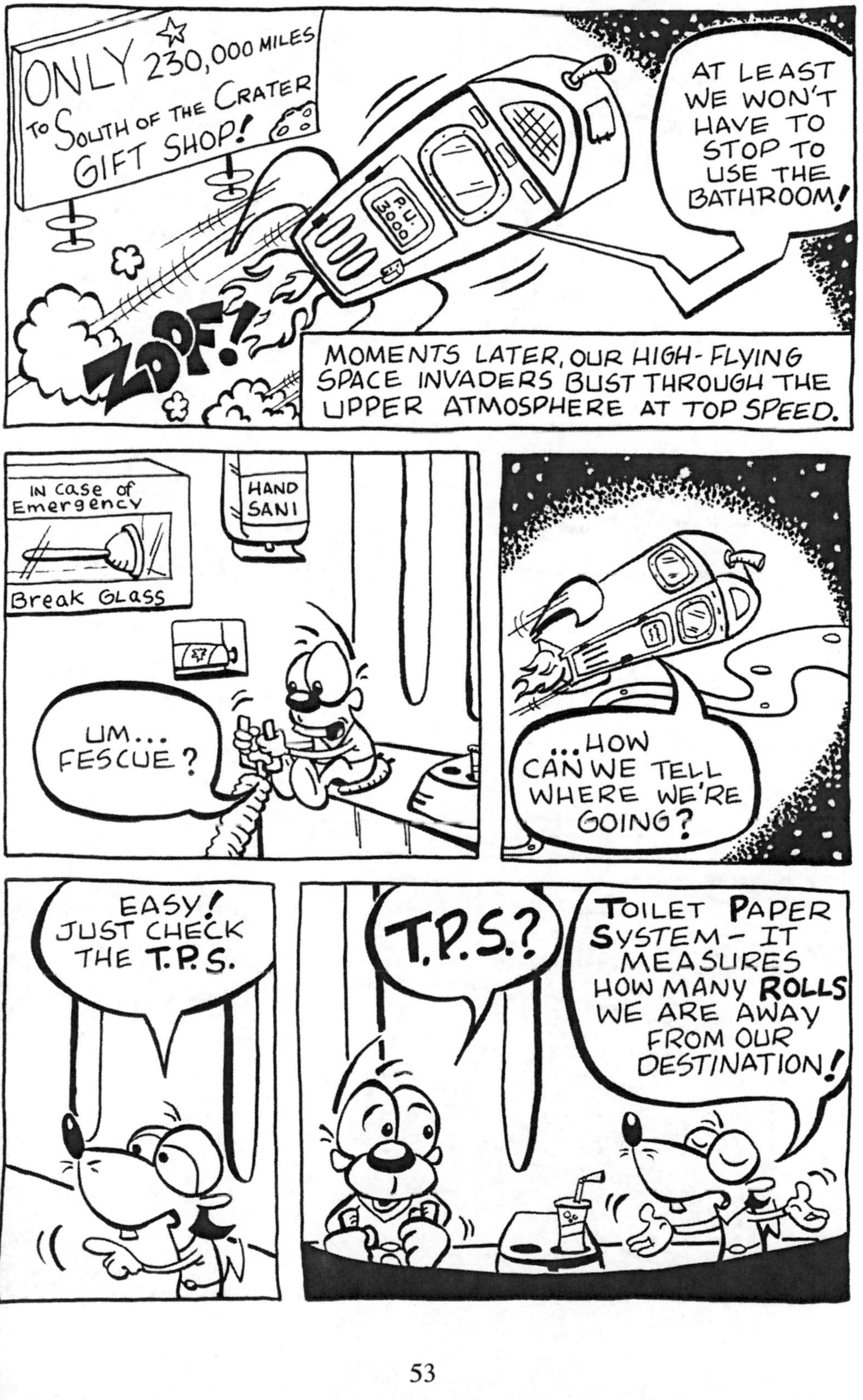
ONLY 230,000 MILES TO SOUTH OF THE CRATER GIFT SHOP!
P.U. 3000
AT LEAST WE WON'T HAVE TO STOP TO USE THE BATHROOM!
ZOOF!
MOMENTS LATER, OUR HIGH-FLYING SPACE INVADERS BUST THROUGH THE UPPER ATMOSPHERE AT TOP SPEED.
IN CASE OF Emergency
Break Glass
HAND SANI
UM... FESCUE?
...HOW CAN WE TELL WHERE WE'RE GOING?
EASY! JUST CHECK THE T.P.S.
T.P.S.?
TOILET PAPER SYSTEM - IT MEASURES HOW MANY ROLLS WE ARE AWAY FROM OUR DESTINATION!

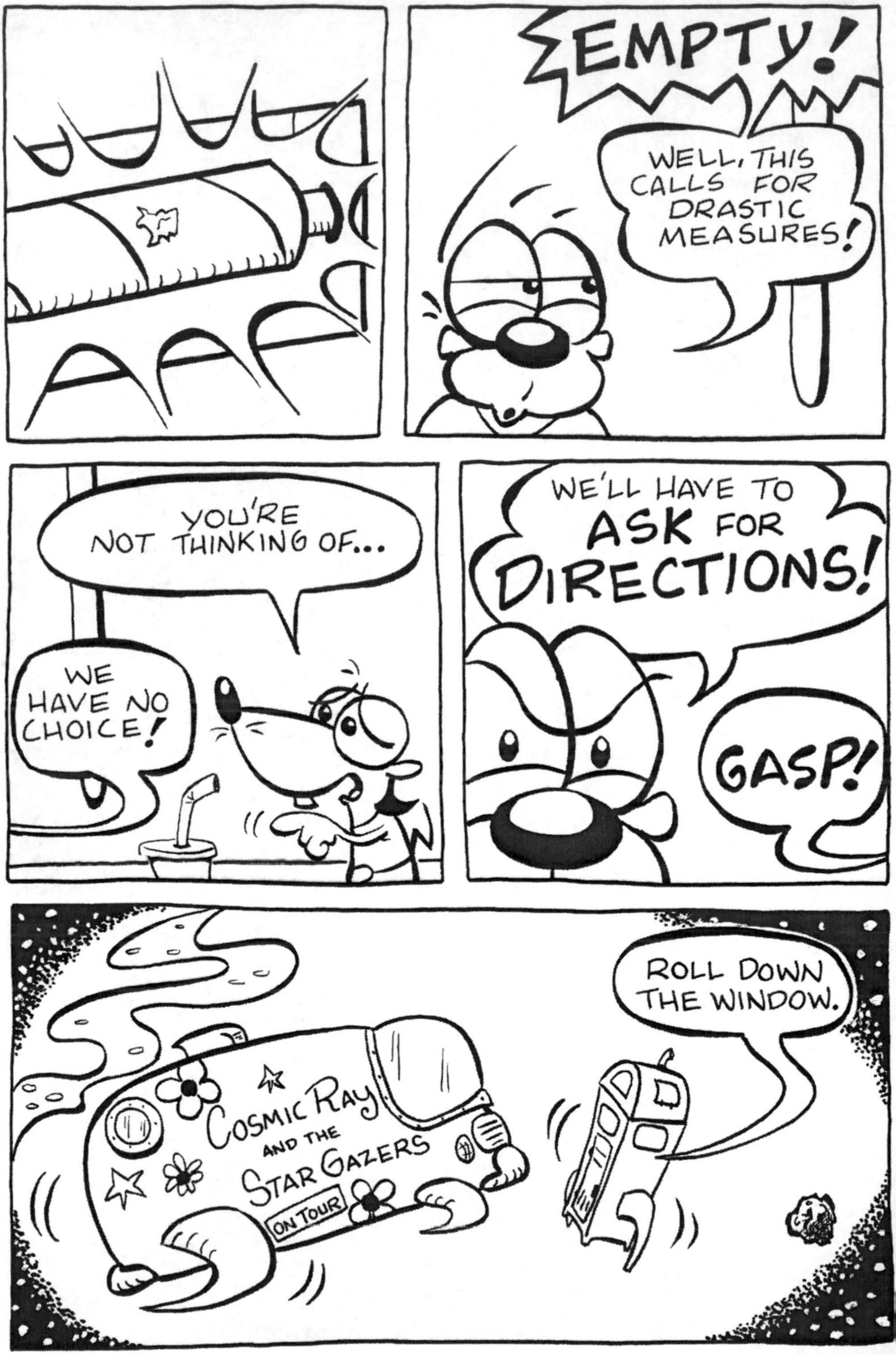
EMPTY!
WELL, THIS CALLS FOR DRASTIC MEASURES!
YOU'RE NOT THINKING OF...
WE HAVE NO CHOICE!
WE'LL HAVE TO ASK FOR DIRECTIONS!
GASP!
ROLL DOWN THE WINDOW.
COSMIC RAY AND THE STAR GAZERS
ON TOUR

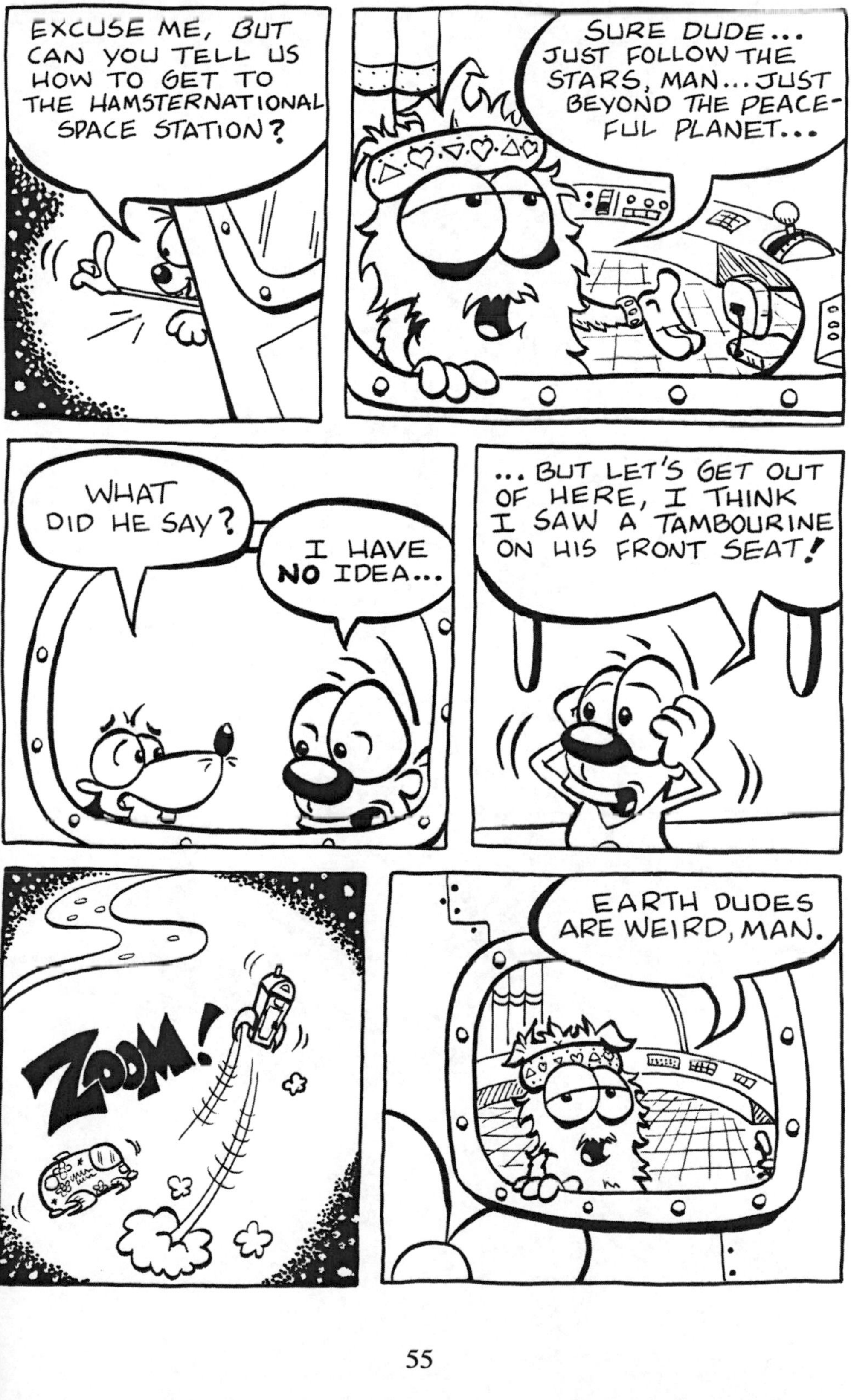
EXCUSE ME, BUT CAN YOU TELL US HOW TO GET TO THE HAMSTERNATIONAL SPACE STATION?
SURE DUDE... JUST FOLLOW THE STARS, MAN... JUST BEYOND THE PEACE-FUL PLANET...
WHAT DID HE SAY?
I HAVE NO IDEA...
... BUT LET'S GET OUT OF HERE, I THINK I SAW A TAMBOURINE ON HIS FRONT SEAT!
ZOOM!
EARTH DUDES ARE WEIRD, MAN.

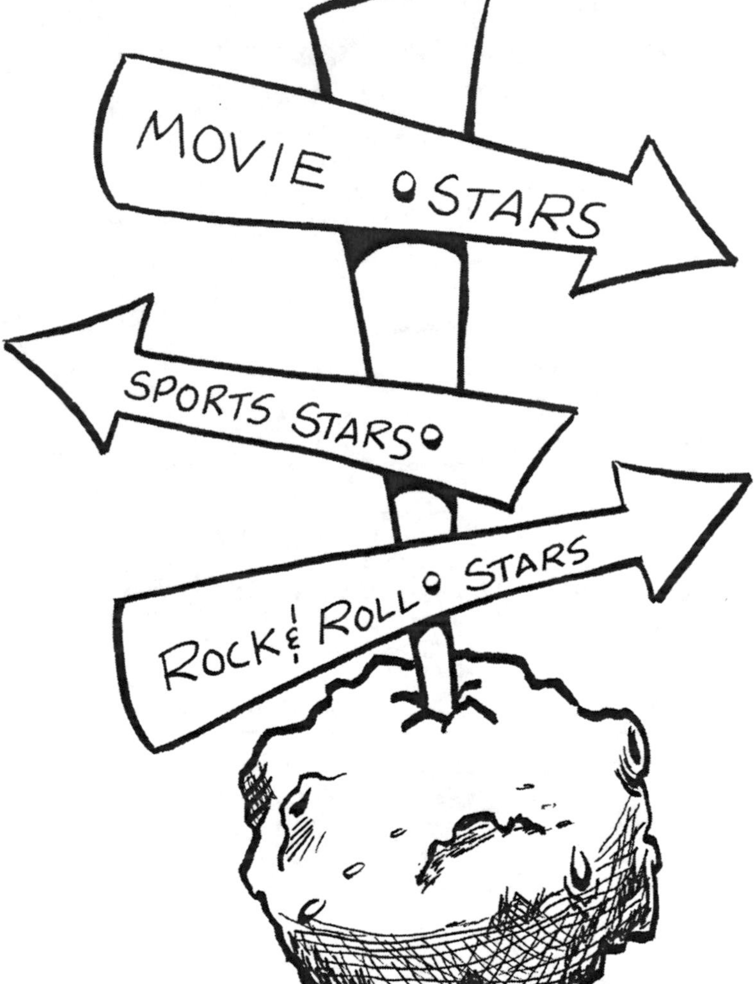
MOVIE STARS
SPORTS STARS
ROCK & ROLL STARS

Chapter Six

A Rather Unsightly
Nose Heir

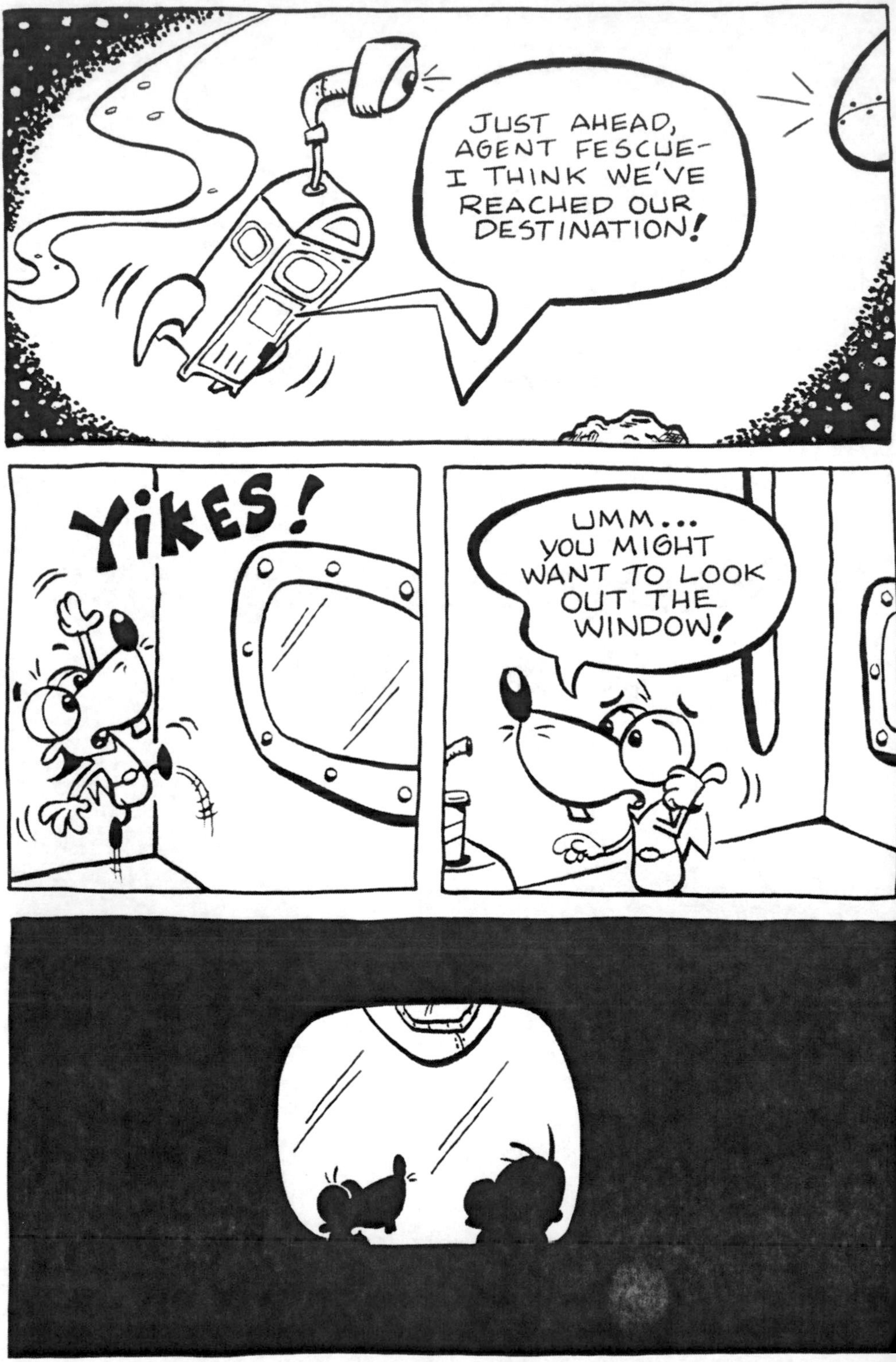
JUST AHEAD, AGENT FESCUE- I THINK WE'VE REACHED OUR DESTINATION!
YIKES!
UMM... YOU MIGHT WANT TO LOOK OUT THE WINDOW!

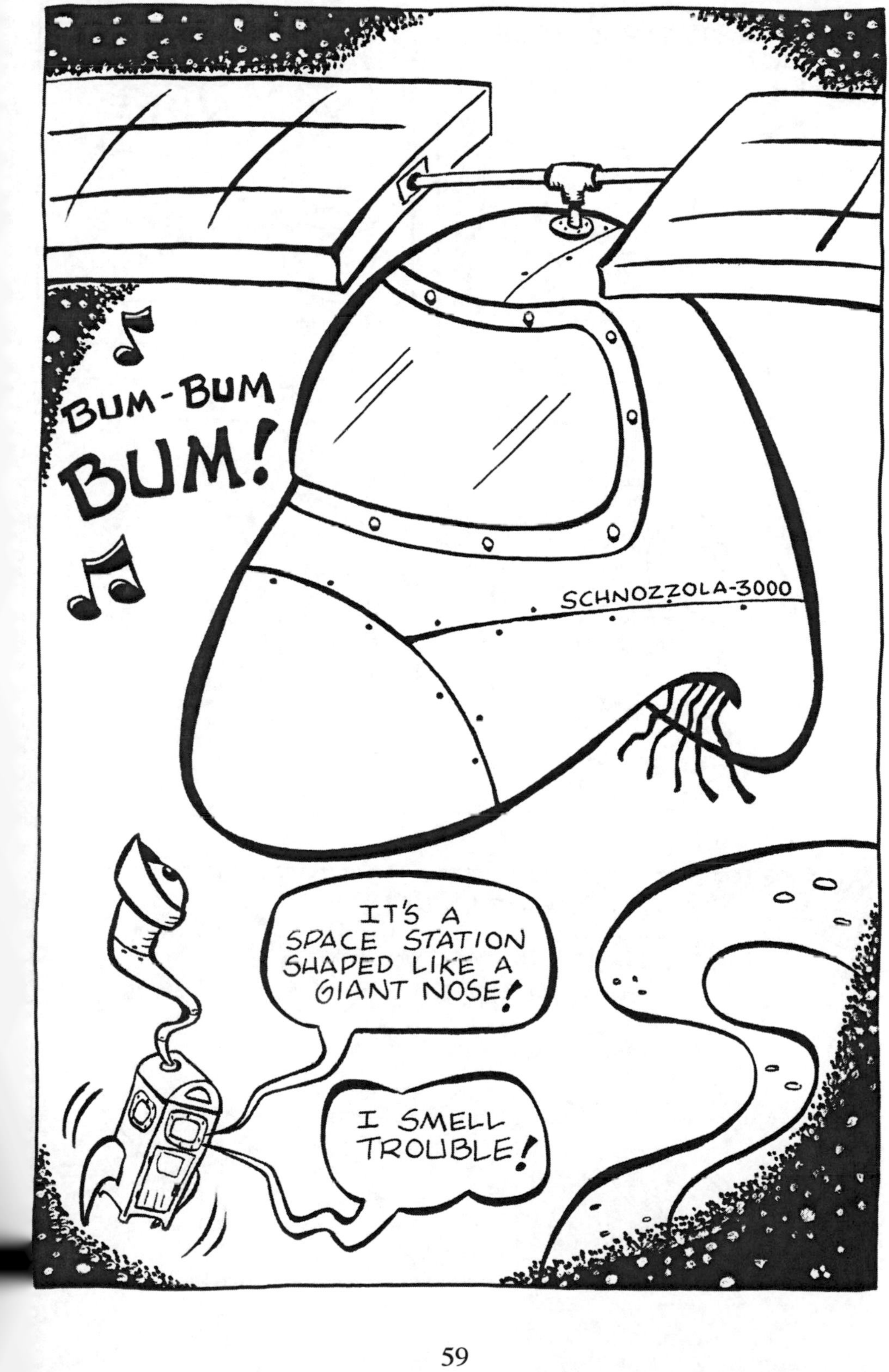
BUM-BUM BUM!
SCHNOZZOLA-3000
IT'S A SPACE STATION SHAPED LIKE A GIANT NOSE!
I SMELL TROUBLE!

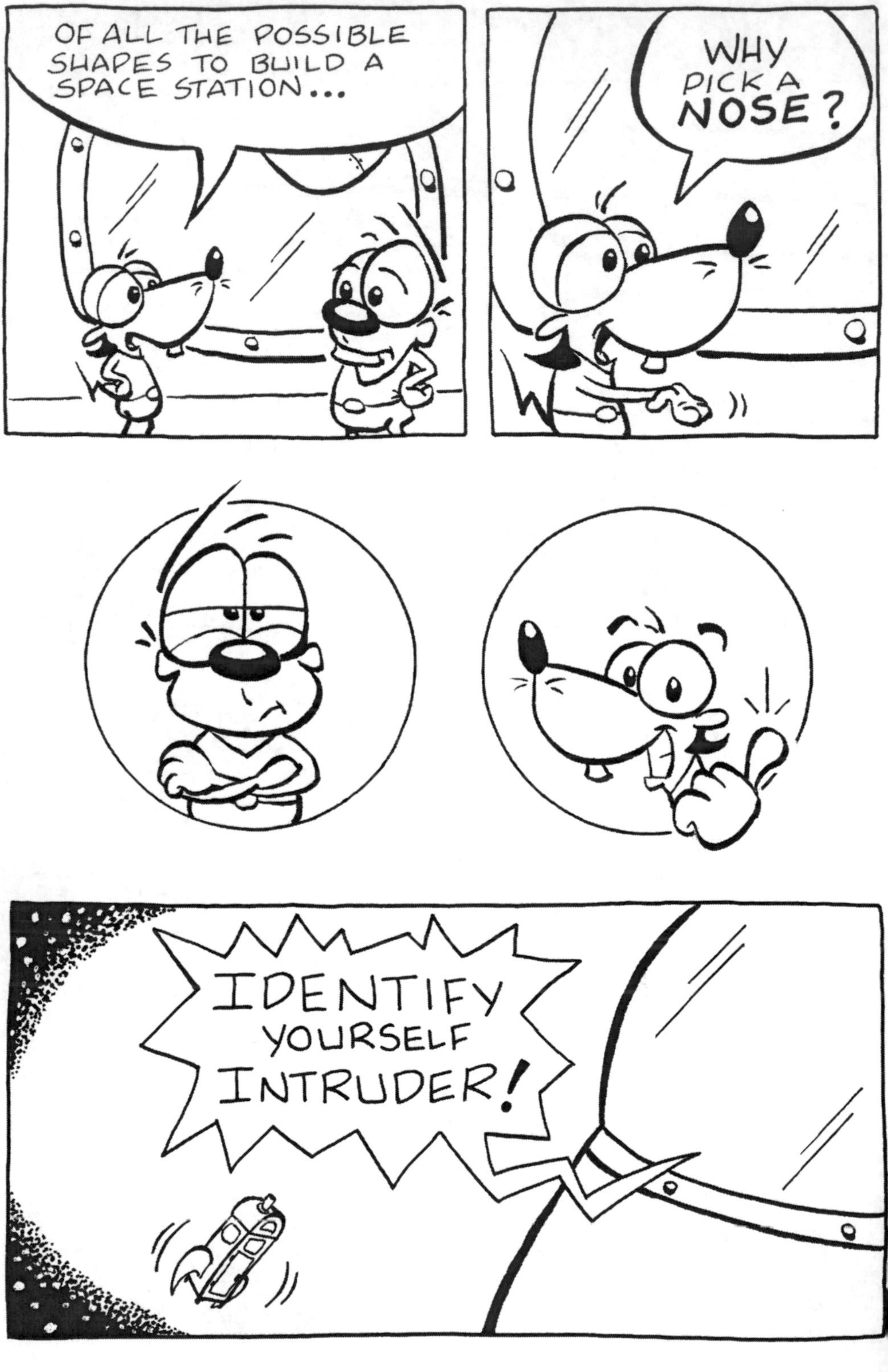
OF ALL THE POSSIBLE SHAPES TO BUILD A SPACE STATION...
WHY PICK A NOSE?
IDENTIFY YOURSELF INTRUDER!

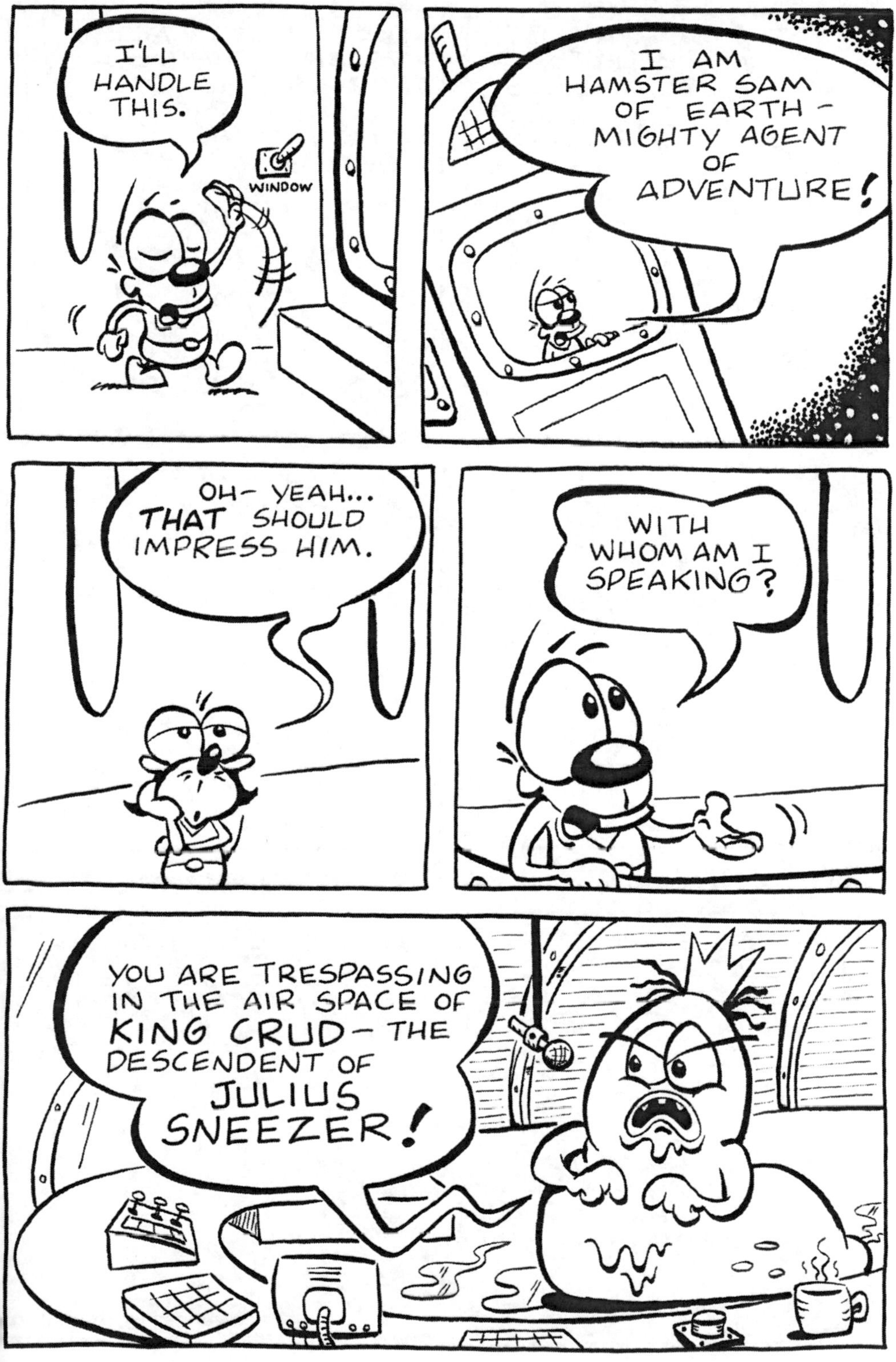
I'LL HANDLE THIS.
WINDOW
I AM HAMSTER SAM OF EARTH - MIGHTY AGENT OF ADVENTURE!
OH- YEAH... THAT SHOULD IMPRESS HIM.
WITH WHOM AM I SPEAKING?
YOU ARE TRESPASSING IN THE AIR SPACE OF KING CRUD- THE DESCENDENT OF JULIUS SNEEZER!

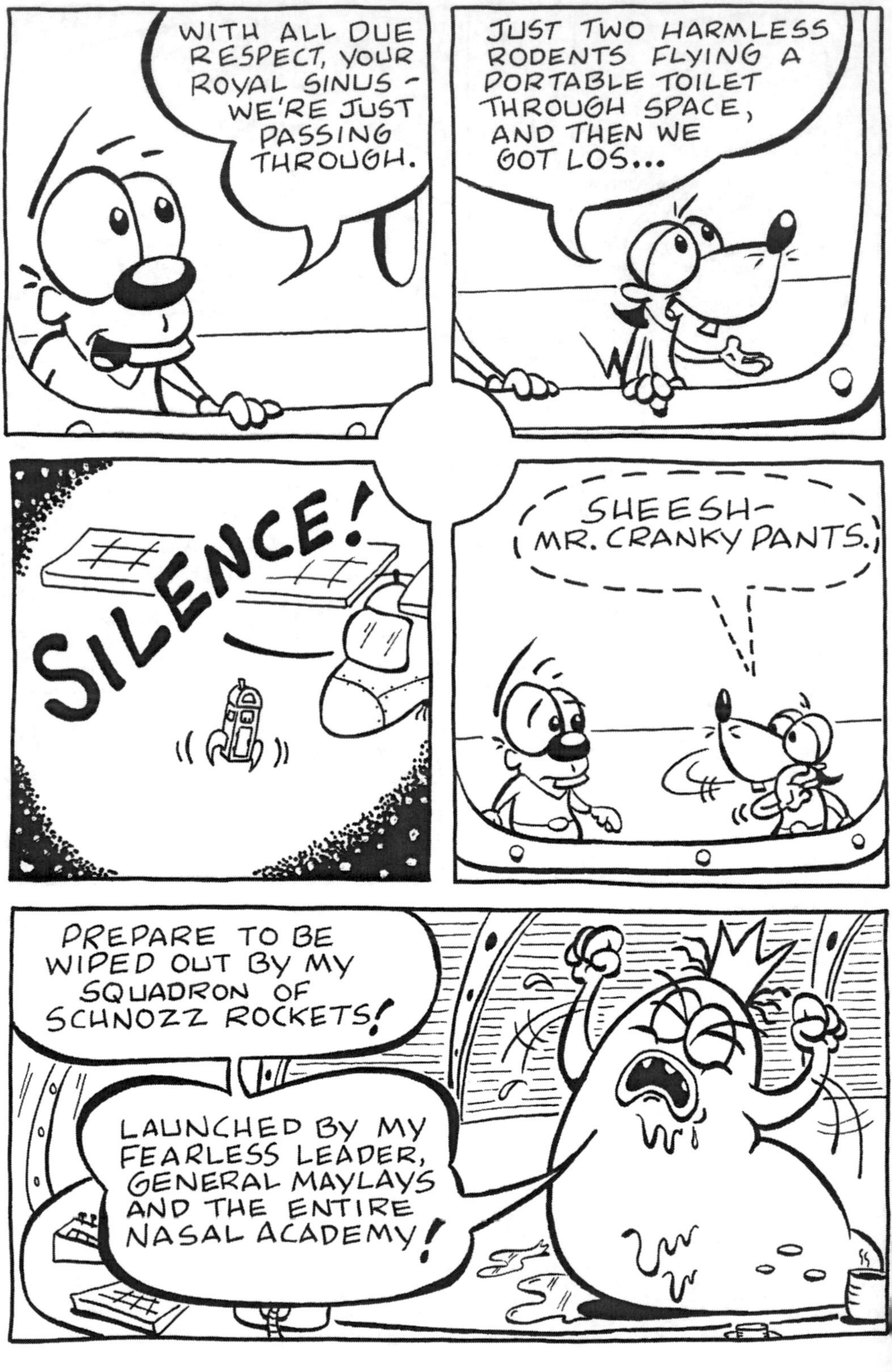
WITH ALL DUE RESPECT, YOUR ROYAL SINUS - WE'RE JUST PASSING THROUGH.
JUST TWO HARMLESS RODENTS FLYING A PORTABLE TOILET THROUGH SPACE, AND THEN WE GOT LOS...
SILENCE!
SHEESH- MR. CRANKY PANTS.
PREPARE TO BE WIPED OUT BY MY SQUADRON OF SCHNOZZ ROCKETS!
LAUNCHED BY MY FEARLESS LEADER, GENERAL MAYLAYS AND THE ENTIRE NASAL ACADEMY!

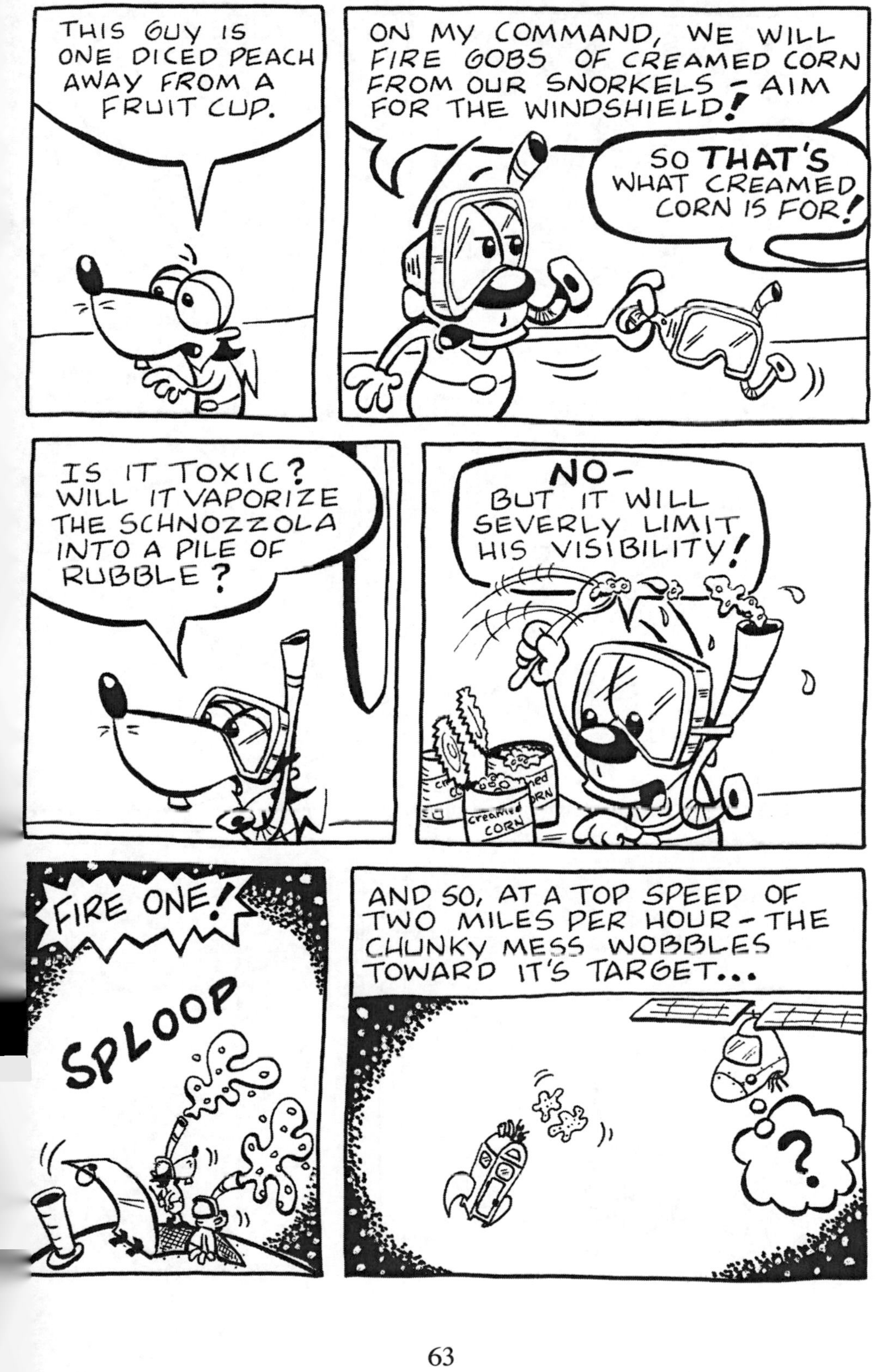
THIS GUY IS ONE DICED PEACH AWAY FROM A FRUIT CUP.
ON MY COMMAND, WE WILL FIRE GOBS OF CREAMED CORN FROM OUR SNORKELS - AIM FOR THE WINDSHIELD!
SO THAT'S WHAT CREAMED CORN IS FOR!
IS IT TOXIC? WILL IT VAPORIZE THE SCHNOZZOLA INTO A PILE OF RUBBLE?
NO- BUT IT WILL SEVERLY LIMIT HIS VISIBILITY!
Creamed CORN
FIRE ONE!
SPLOOP
AND SO, AT A TOP SPEED OF TWO MILES PER HOUR - THE CHUNKY MESS WOBBLES TOWARD IT'S TARGET...
?

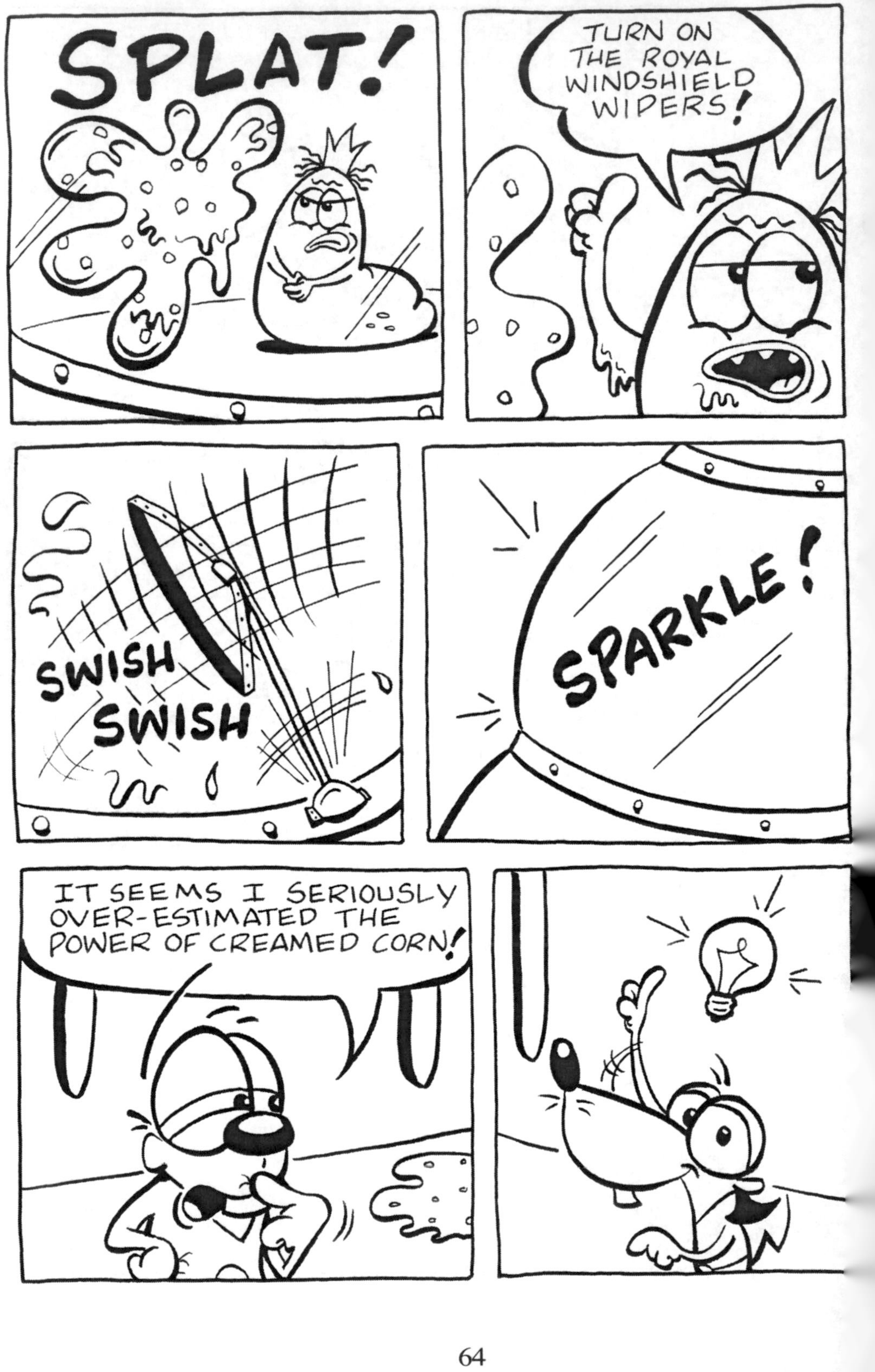
SPLAT!
TURN ON THE ROYAL WINDSHIELD WIPERS!
SWISH
SWISH
SPARKLE!
IT SEEMS I SERIOUSLY OVER-ESTIMATED THE POWER OF CREAMED CORN!

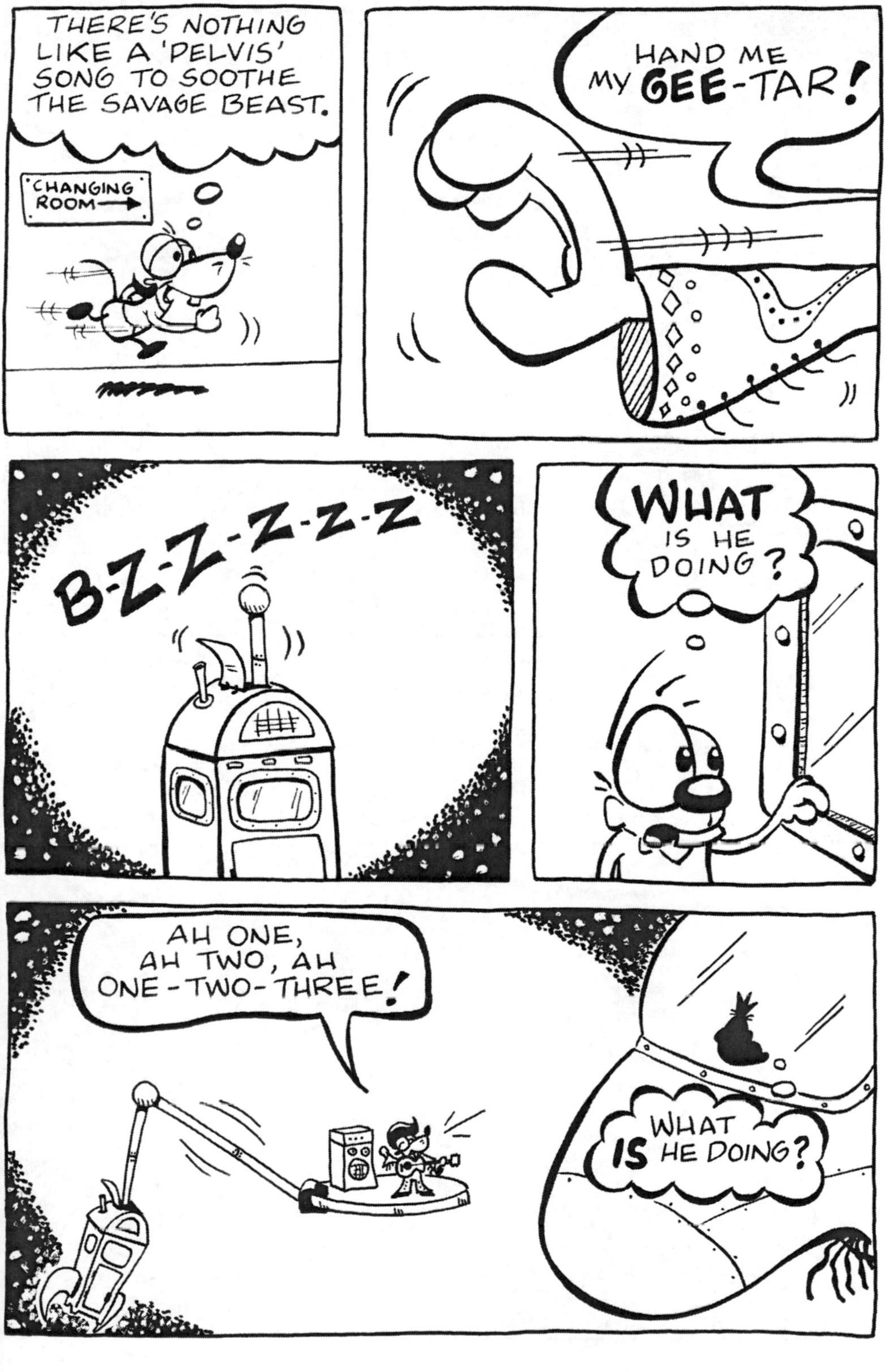
THERE'S NOTHING LIKE A 'PELVIS' SONG TO SOOTHE THE SAVAGE BEAST.
CHANGING ROOM
HAND ME MY GEE-TAR!
B-Z-Z-Z-Z-Z
WHAT IS HE DOING?
AH ONE, AH TWO, AH ONE-TWO-THREE!
WHAT IS HE DOING?

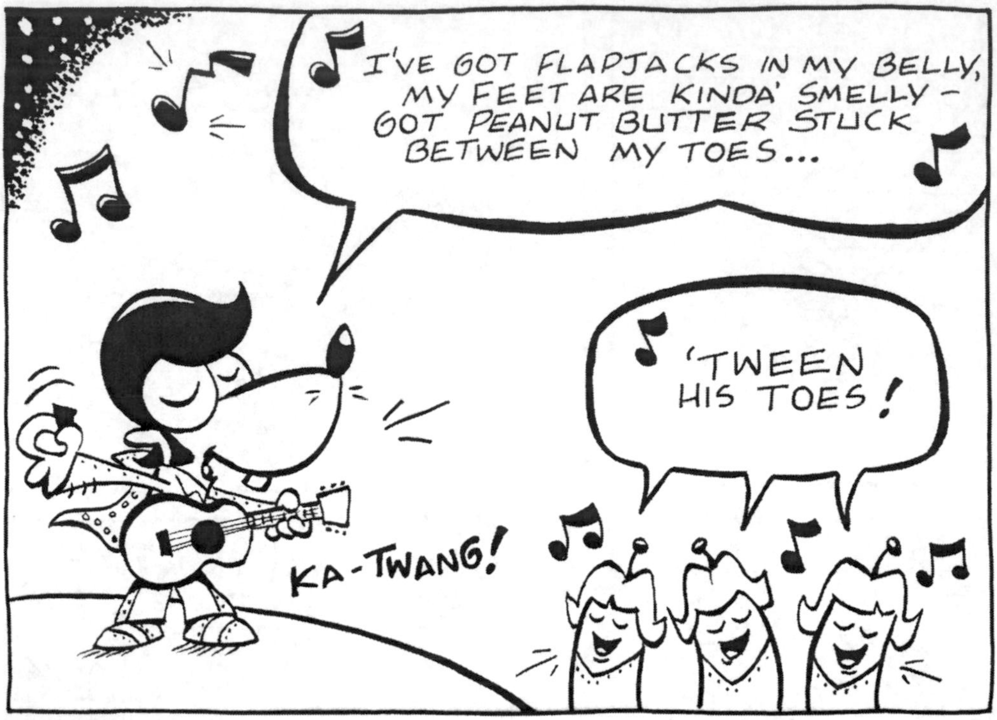
I'VE GOT FLAPJACKS IN MY BELLY, MY FEET ARE KINDA' SMELLY - GOT PEANUT BUTTER STUCK BETWEEN MY TOES...
'TWEEN HIS TOES!
KA-TWANG!

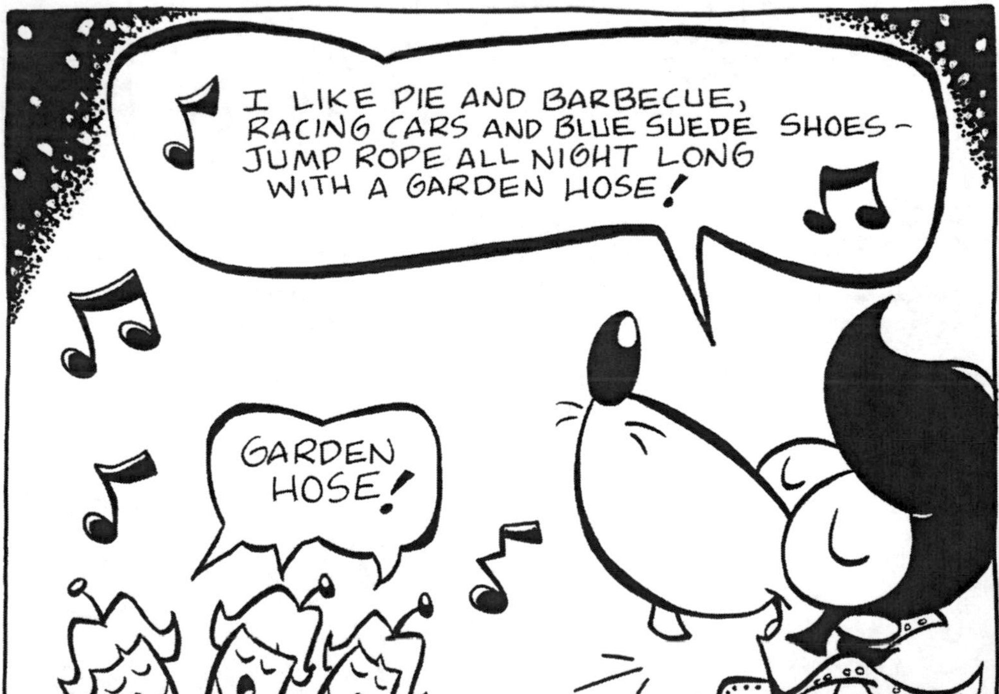
I LIKE PIE AND BARBECUE, RACING CARS AND BLUE SUEDE SHOES - JUMP ROPE ALL NIGHT LONG WITH A GARDEN HOSE!
GARDEN HOSE!

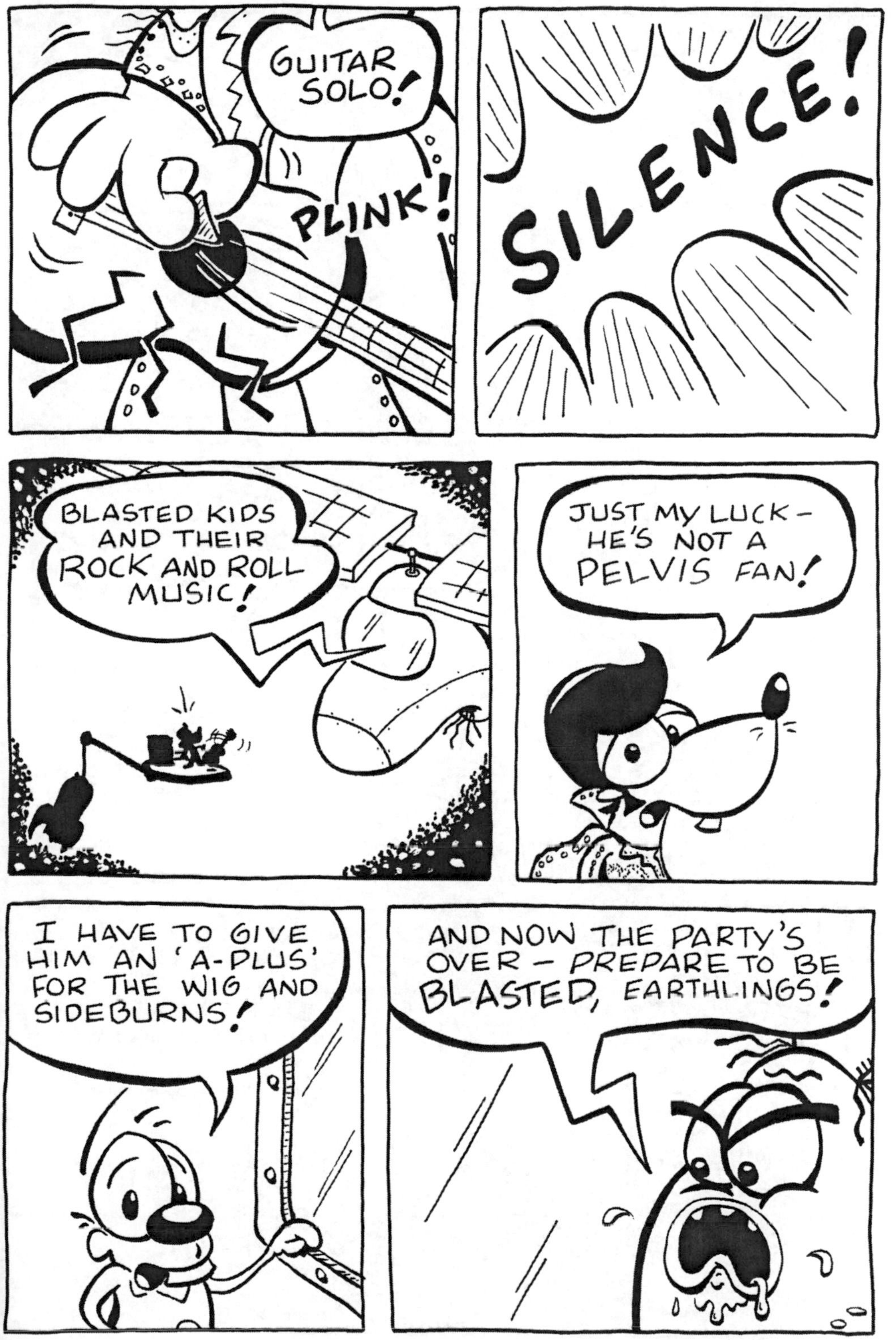
GUITAR SOLO!
PLINK!
SILENCE!
BLASTED KIDS AND THEIR ROCK AND ROLL MUSIC!
JUST MY LUCK- HE'S NOT A PELVIS FAN!
I HAVE TO GIVE HIM AN 'A-PLUS' FOR THE WIG AND SIDEBURNS!
AND NOW THE PARTY'S OVER - *PREPARE* TO BE BLASTED, EARTHLINGS!

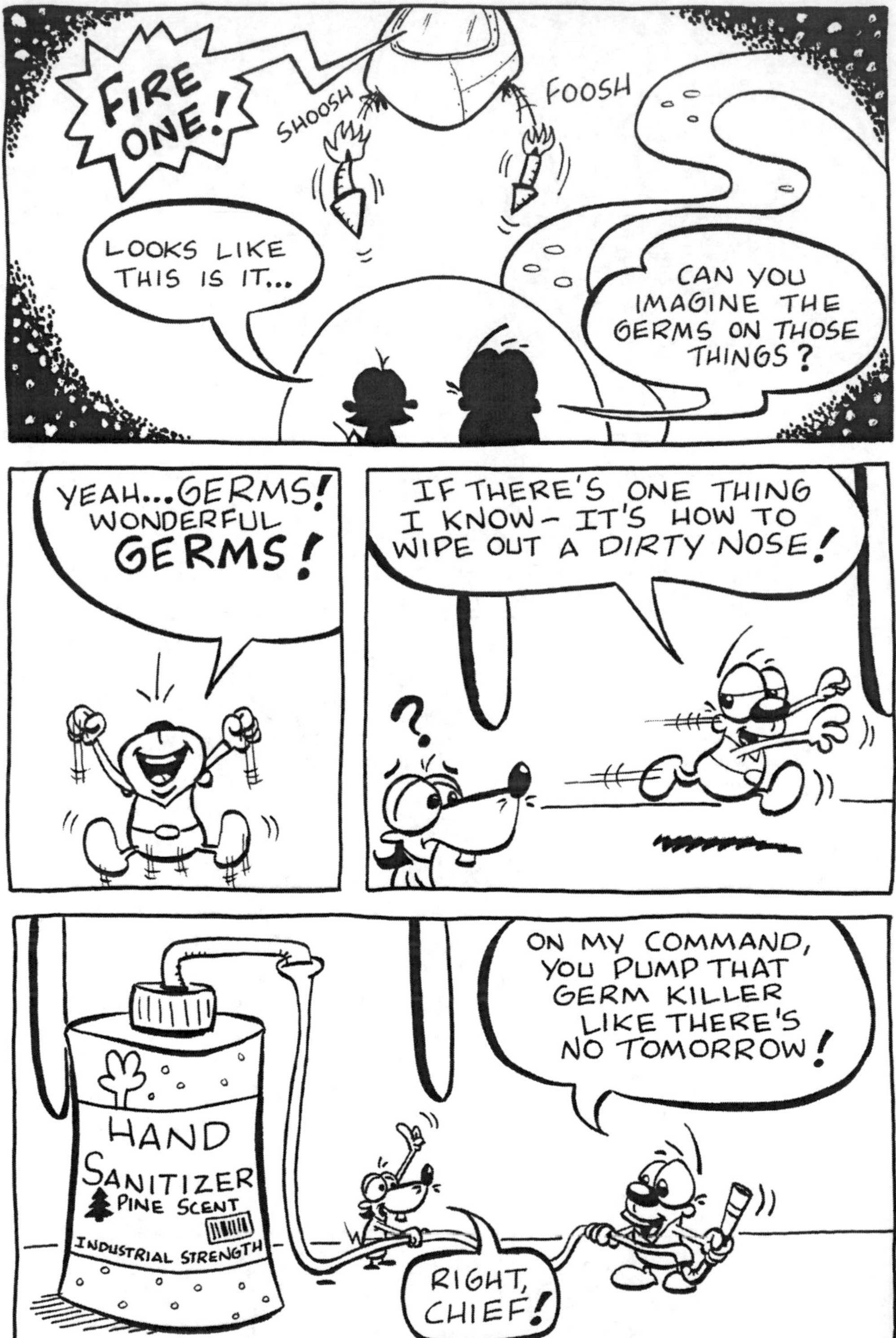
FIRE ONE!
SHOOSH
FOOSH
LOOKS LIKE THIS IS IT...
CAN YOU IMAGINE THE GERMS ON THOSE THINGS?
YEAH...GERMS! WONDERFUL GERMS!
IF THERE'S ONE THING I KNOW- IT'S HOW TO WIPE OUT A DIRTY NOSE!
?
HAND SANITIZER
PINE SCENT
INDUSTRIAL STRENGTH
ON MY COMMAND, YOU PUMP THAT GERM KILLER LIKE THERE'S NO TOMORROW!
RIGHT, CHIEF!

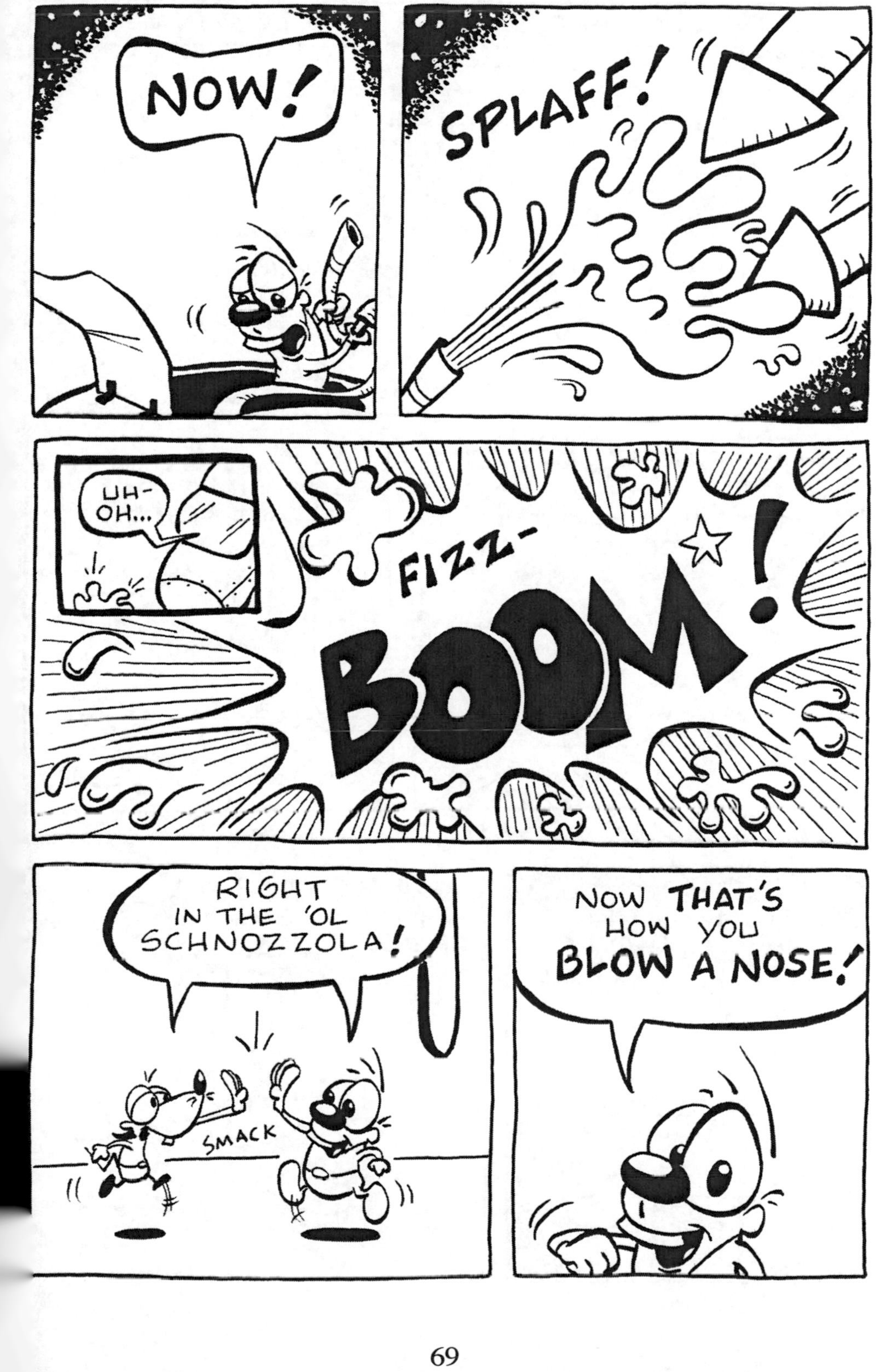
NOW!
SPLAFF!
UH-OH...
FIZZ-
BOOM!
RIGHT IN THE 'OL SCHNOZZOLA!
SMACK
NOW THAT'S HOW YOU BLOW A NOSE!

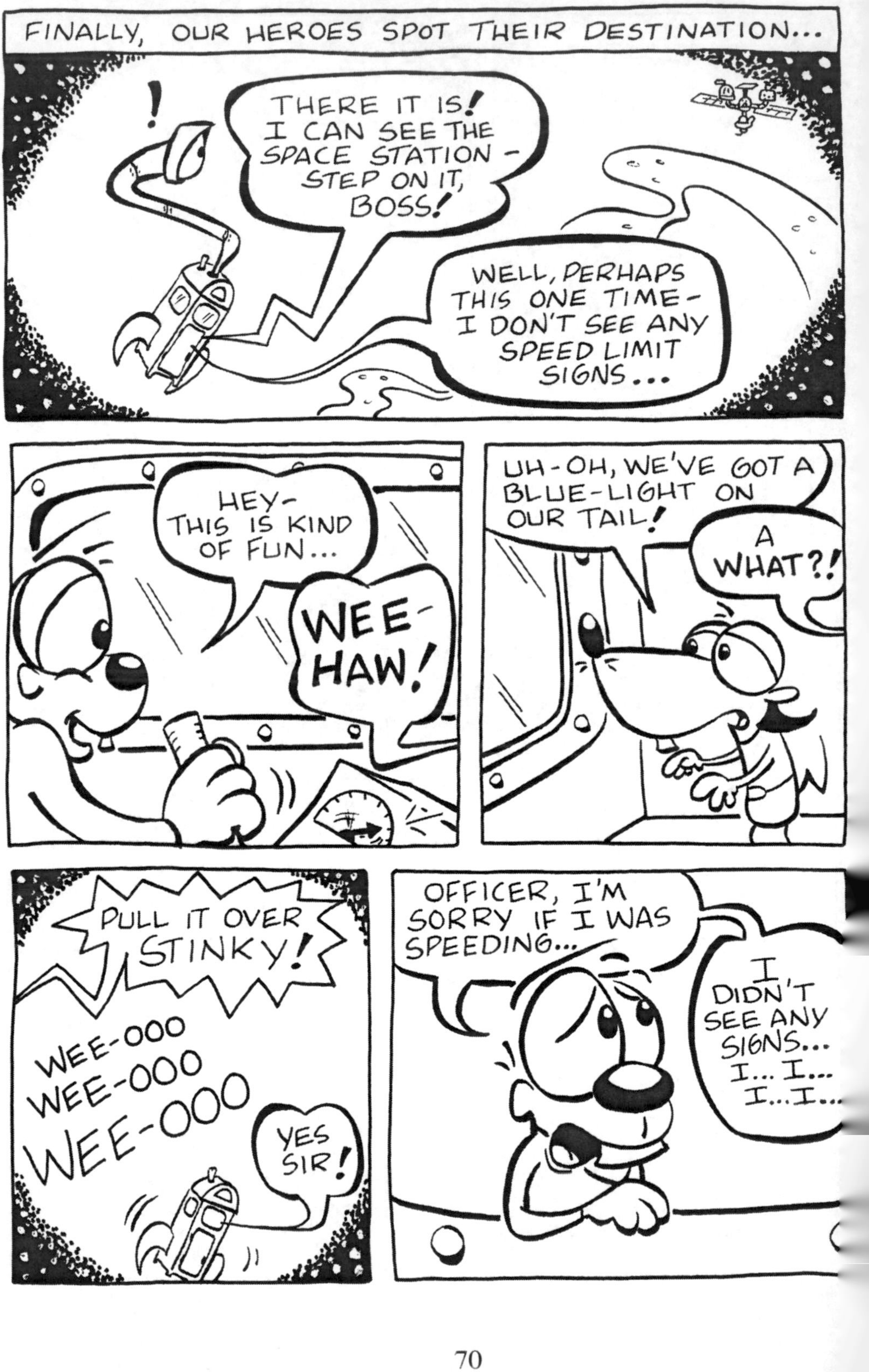
FINALLY, OUR HEROES SPOT THEIR DESTINATION...
!
THERE IT IS! I CAN SEE THE SPACE STATION - STEP ON IT, BOSS!
WELL, PERHAPS THIS ONE TIME - I DON'T SEE ANY SPEED LIMIT SIGNS...
HEY- THIS IS KIND OF FUN...
WEE-HAW!
UH-OH, WE'VE GOT A BLUE-LIGHT ON OUR TAIL!
A WHAT?!
PULL IT OVER STINKY!
WEE-OOO
WEE-OOO
WEE-OOO
YES SIR!
OFFICER, I'M SORRY IF I WAS SPEEDING...
I DIDN'T SEE ANY SIGNS... I... I... I... I...

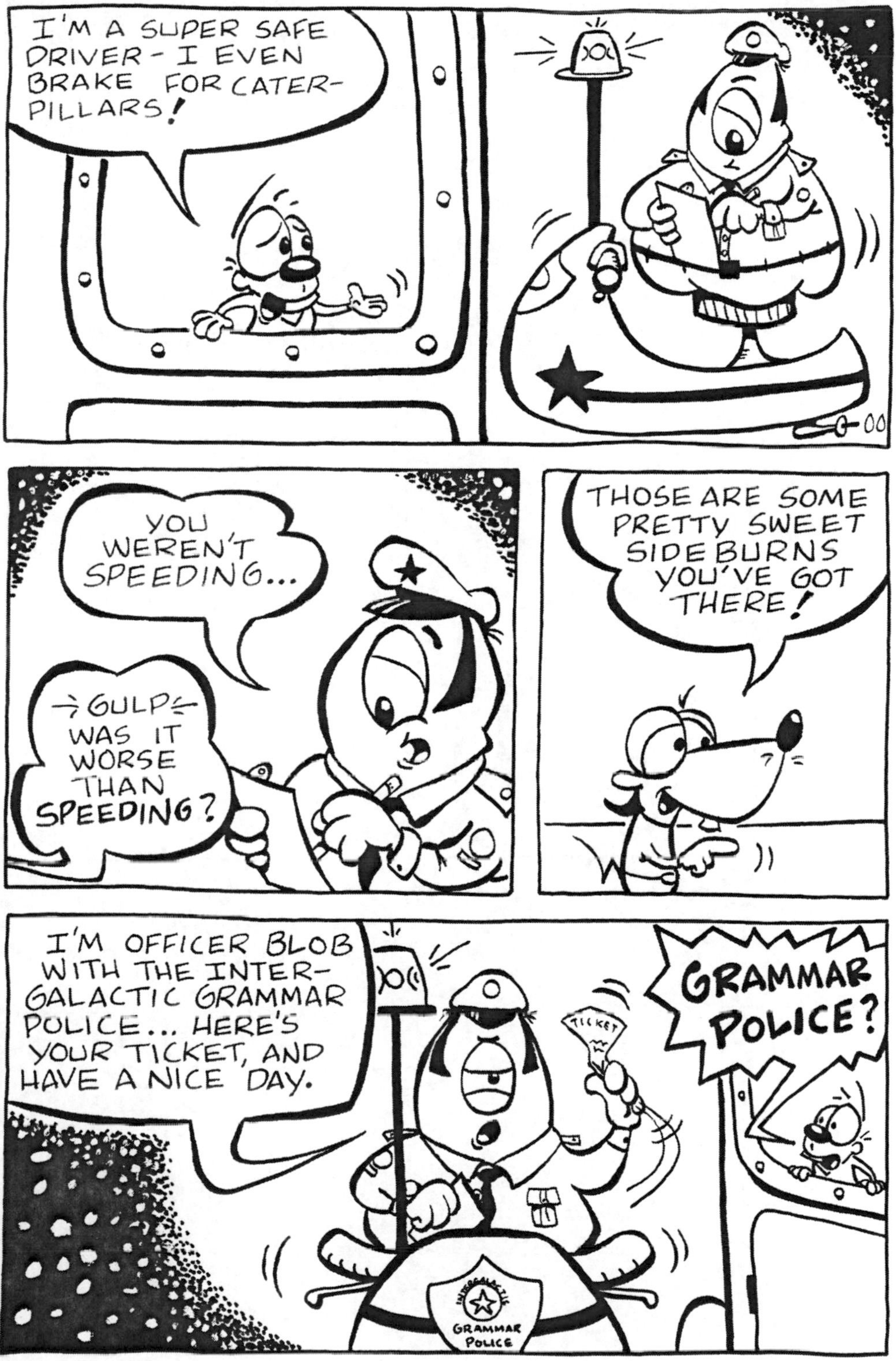
I'M A SUPER SAFE DRIVER - I EVEN BRAKE FOR CATER-PILLARS!
YOU WEREN'T SPEEDING...
GULP
WAS IT WORSE THAN SPEEDING?
THOSE ARE SOME PRETTY SWEET SIDEBURNS YOU'VE GOT THERE!
I'M OFFICER BLOB WITH THE INTER-GALACTIC GRAMMAR POLICE... HERE'S YOUR TICKET, AND HAVE A NICE DAY.
TICKET
GRAMMAR POLICE?
INTERGALACTIC GRAMMAR POLICE

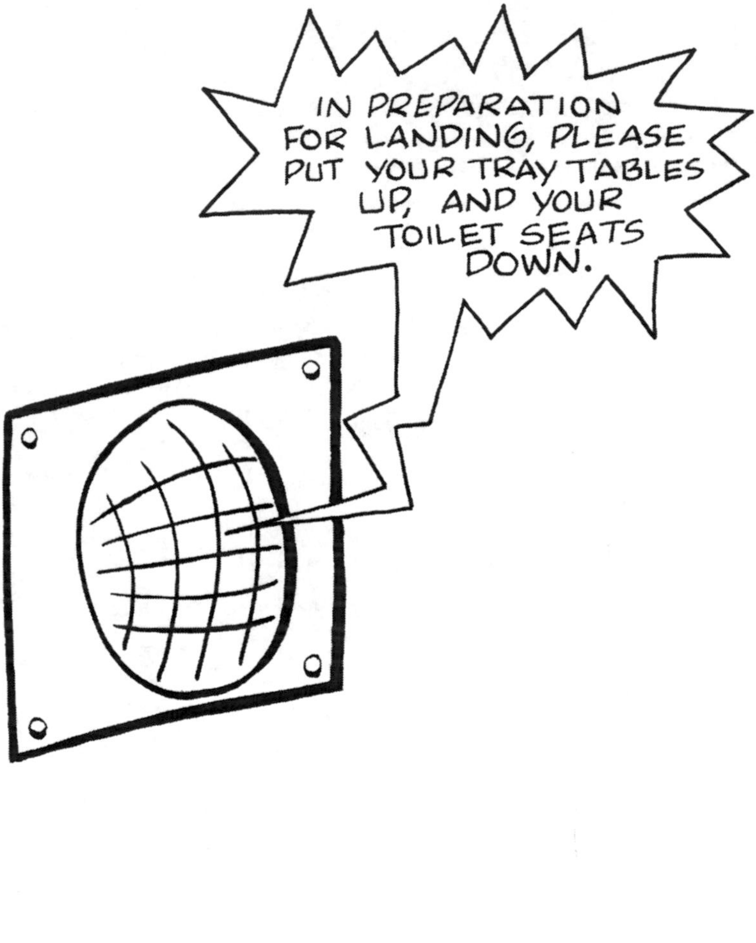
IN PREPARATION FOR LANDING, PLEASE PUT YOUR TRAY TABLES UP, AND YOUR TOILET SEATS DOWN.

Chapter Seven

You Have Reached Your DUST-ination!

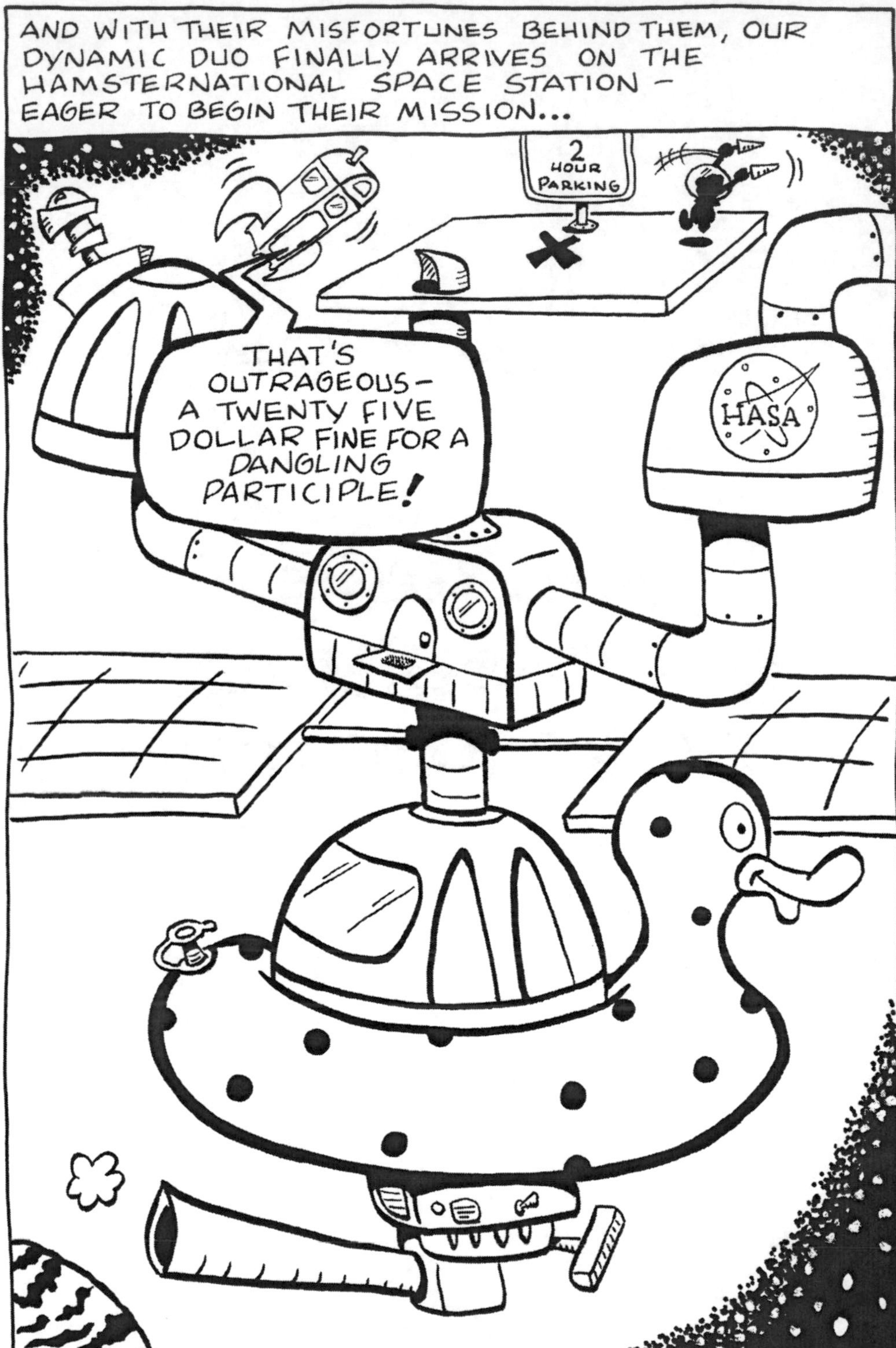
AND WITH THEIR MISFORTUNES BEHIND THEM, OUR DYNAMIC DUO FINALLY ARRIVES ON THE HAMSTERNATIONAL SPACE STATION — EAGER TO BEGIN THEIR MISSION...
2 HOUR PARKING
THAT'S OUTRAGEOUS — A TWENTY FIVE DOLLAR FINE FOR A DANGLING PARTICIPLE!
HASA

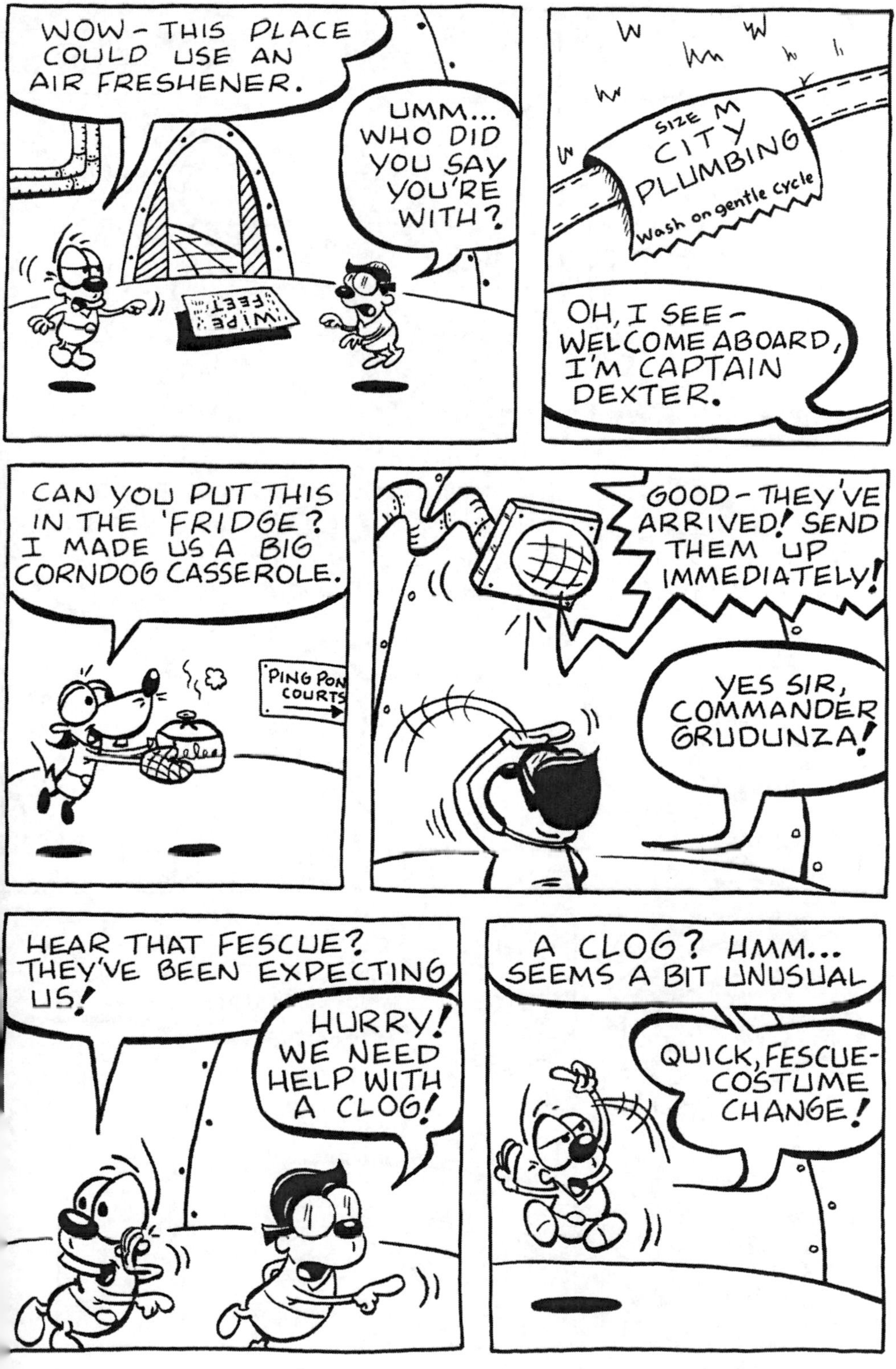
WOW - THIS PLACE COULD USE AN AIR FRESHENER.
WIPE FEET
UMM... WHO DID YOU SAY YOU'RE WITH?
SIZE M
CITY PLUMBING
Wash on gentle cycle
OH, I SEE - WELCOME ABOARD, I'M CAPTAIN DEXTER.
CAN YOU PUT THIS IN THE 'FRIDGE? I MADE US A BIG CORNDOG CASSEROLE.
PING PON COURTS
GOOD - THEY'VE ARRIVED! SEND THEM UP IMMEDIATELY!
YES SIR, COMMANDER GRUDUNZA!
HEAR THAT FESCUE? THEY'VE BEEN EXPECTING US!
HURRY! WE NEED HELP WITH A CLOG!
A CLOG? HMM... SEEMS A BIT UNUSUAL
QUICK, FESCUE - COSTUME CHANGE!

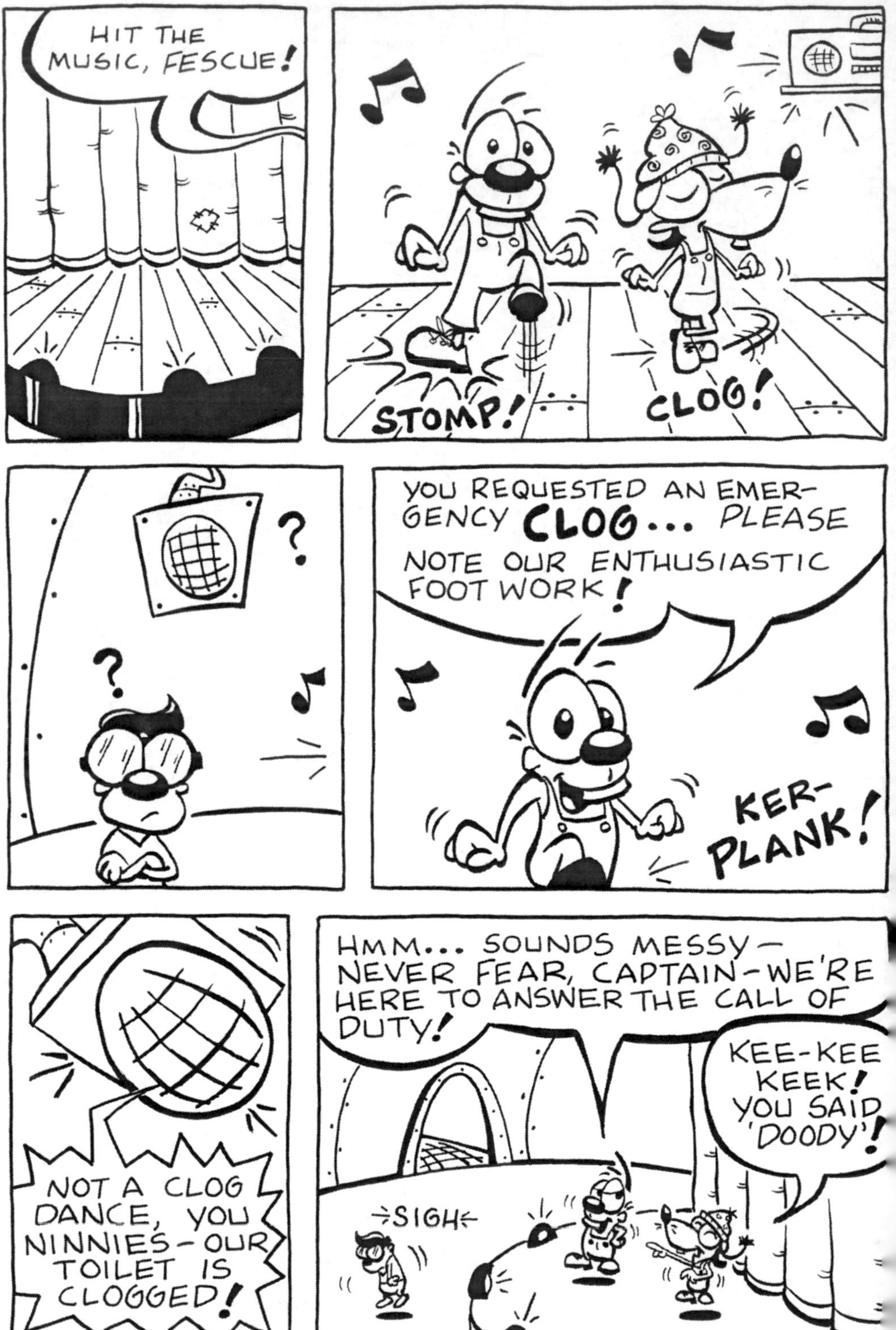
HIT THE MUSIC, FESCUE!
STOMP!
CLOG!
YOU REQUESTED AN EMERGENCY CLOG... PLEASE NOTE OUR ENTHUSIASTIC FOOT WORK!
KER-PLANK!
NOT A CLOG DANCE, YOU NINNIES-OUR TOILET IS CLOGGED!
HMM... SOUNDS MESSY-NEVER FEAR, CAPTAIN-WE'RE HERE TO ANSWER THE CALL OF DUTY!
KEE-KEE KEEK! YOU SAID 'DOODY'!
SIGH

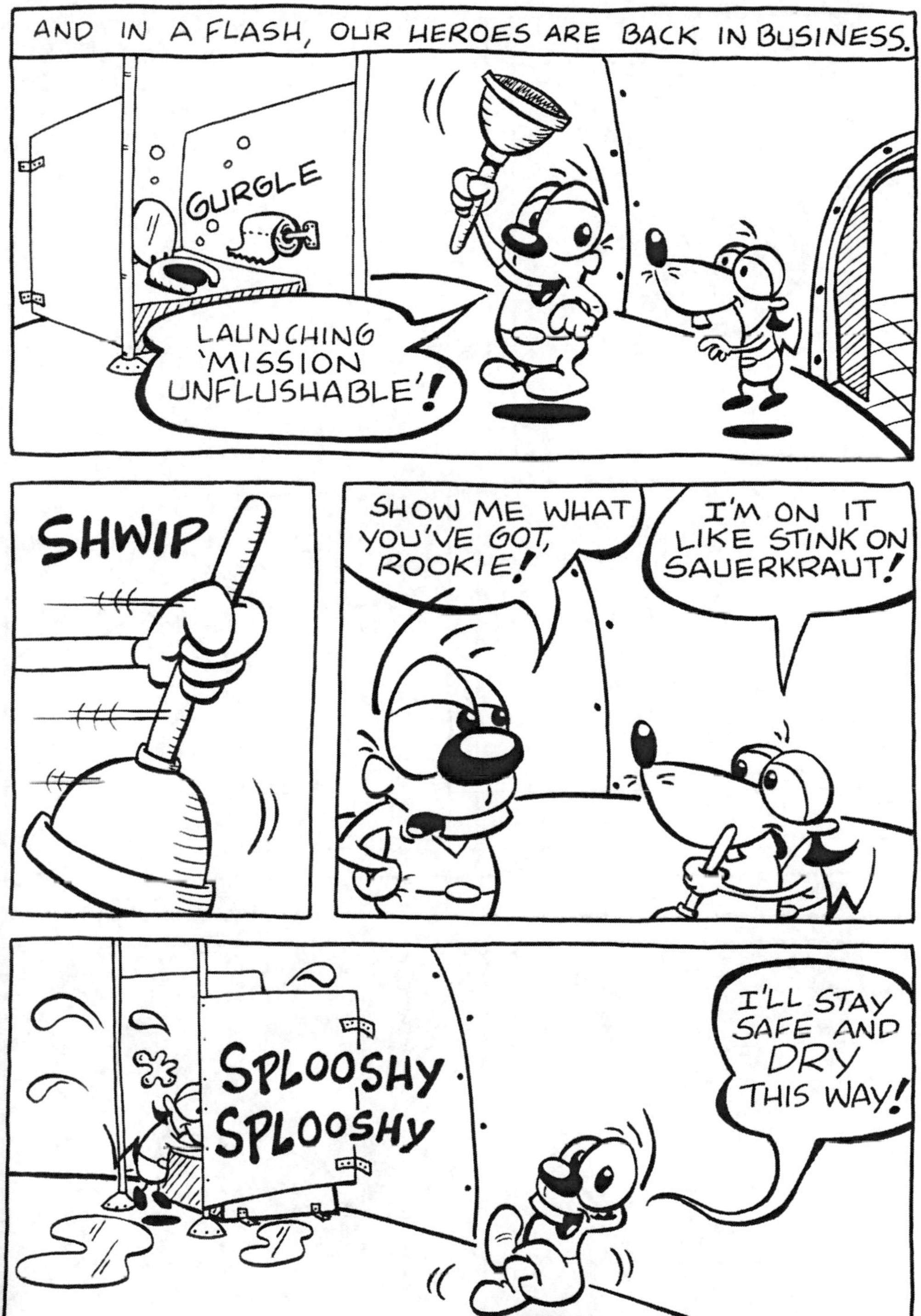
AND IN A FLASH, OUR HEROES ARE BACK IN BUSINESS.
GURGLE
LAUNCHING 'MISSION UNFLUSHABLE'!
SHWIP
SHOW ME WHAT YOU'VE GOT, ROOKIE!
I'M ON IT LIKE STINK ON SAUERKRAUT!
SPLOOSHY SPLOOSHY
I'LL STAY SAFE AND DRY THIS WAY!

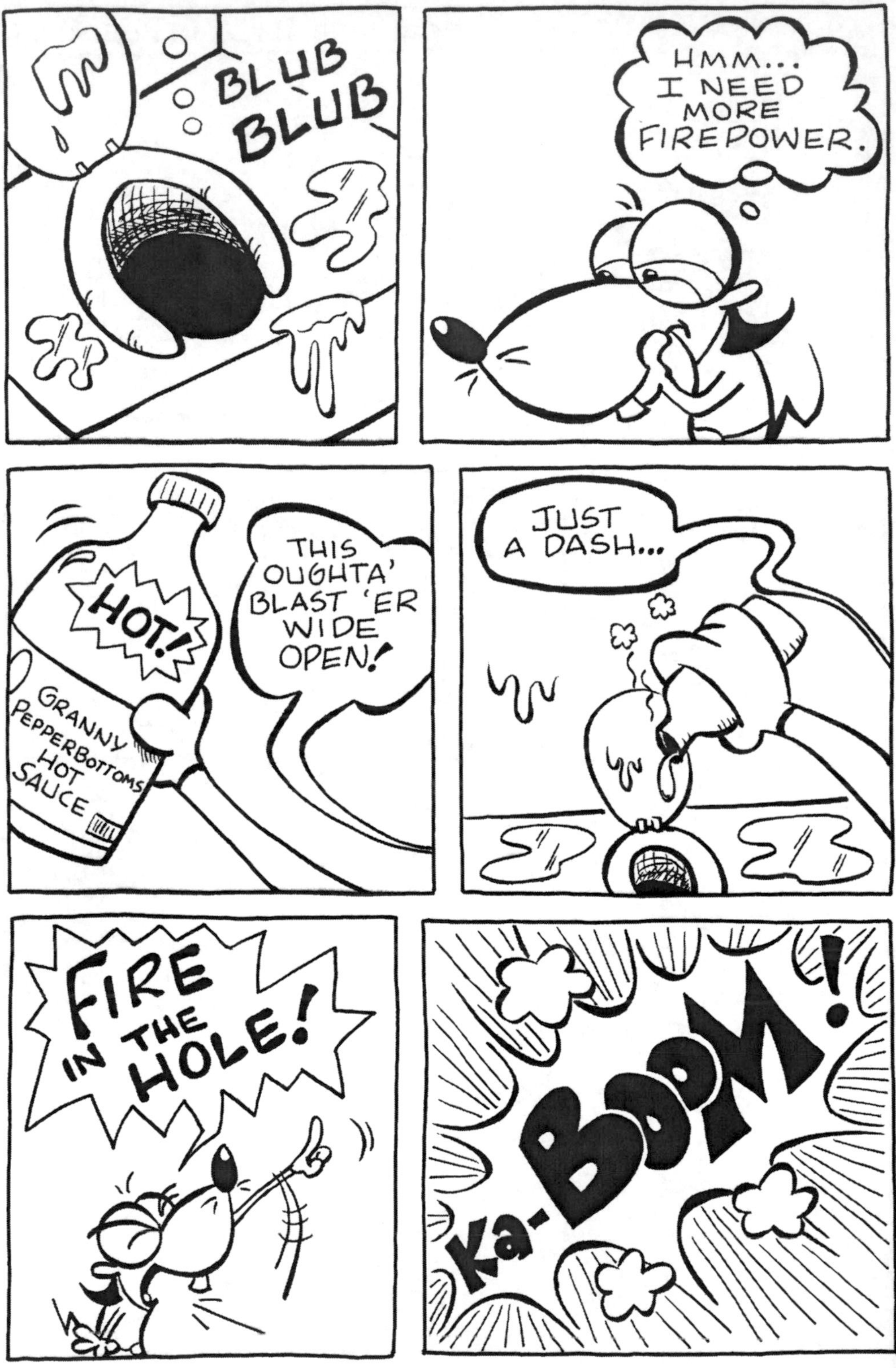
BLUB BLUB
HMM... I NEED MORE FIREPOWER.
HOT!
GRANNY PEPPERBOTTOMS HOT SAUCE
THIS OUGHTA' BLAST 'ER WIDE OPEN!
JUST A DASH...
FIRE IN THE HOLE!
KA-BOOM!

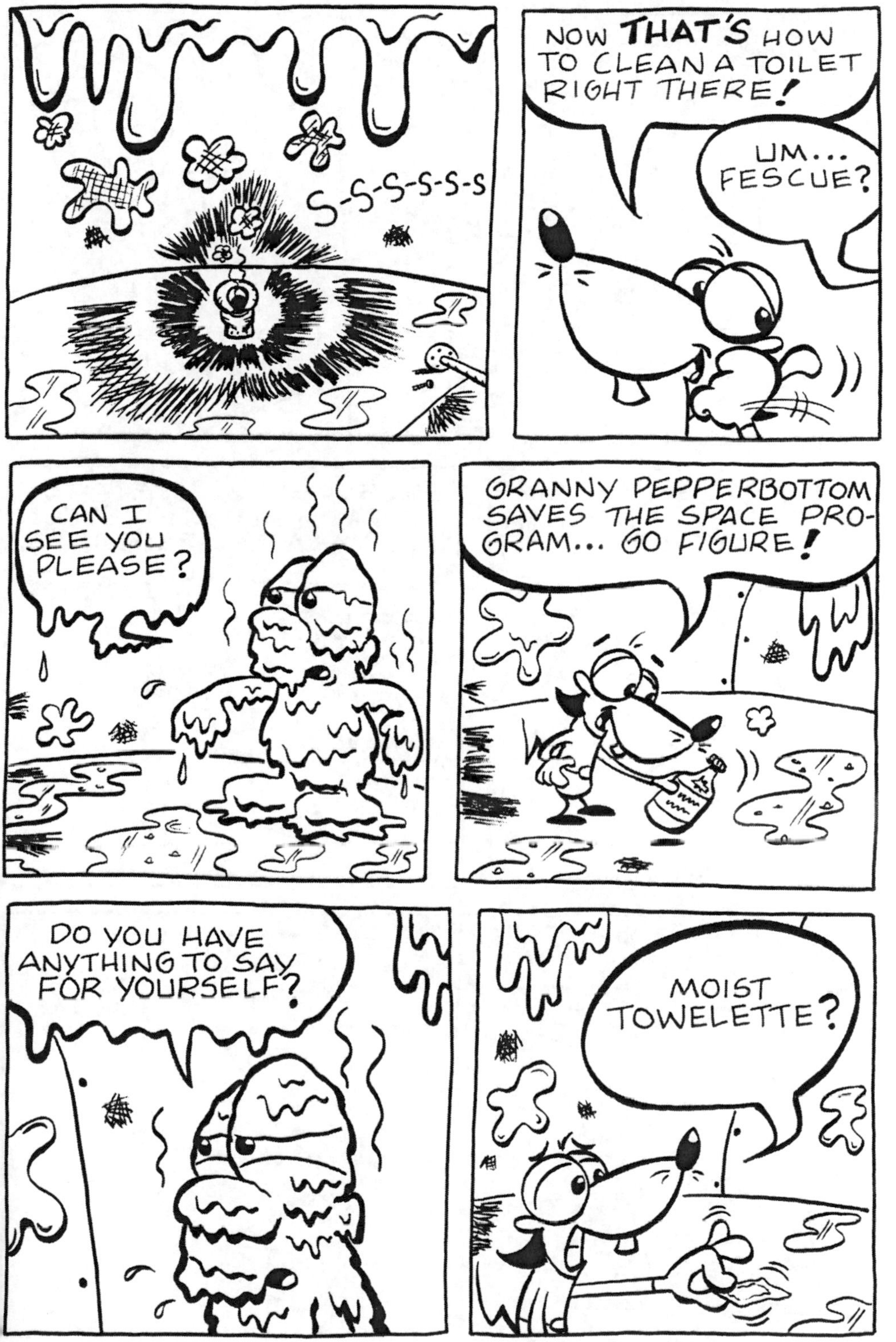
S-S-S-S-S-S
NOW THAT'S HOW TO CLEAN A TOILET RIGHT THERE!
UM... FESCUE?
CAN I SEE YOU PLEASE?
GRANNY PEPPERBOTTOM SAVES THE SPACE PRO-GRAM... GO FIGURE!
DO YOU HAVE ANYTHING TO SAY FOR YOURSELF?
MOIST TOWELETTE?

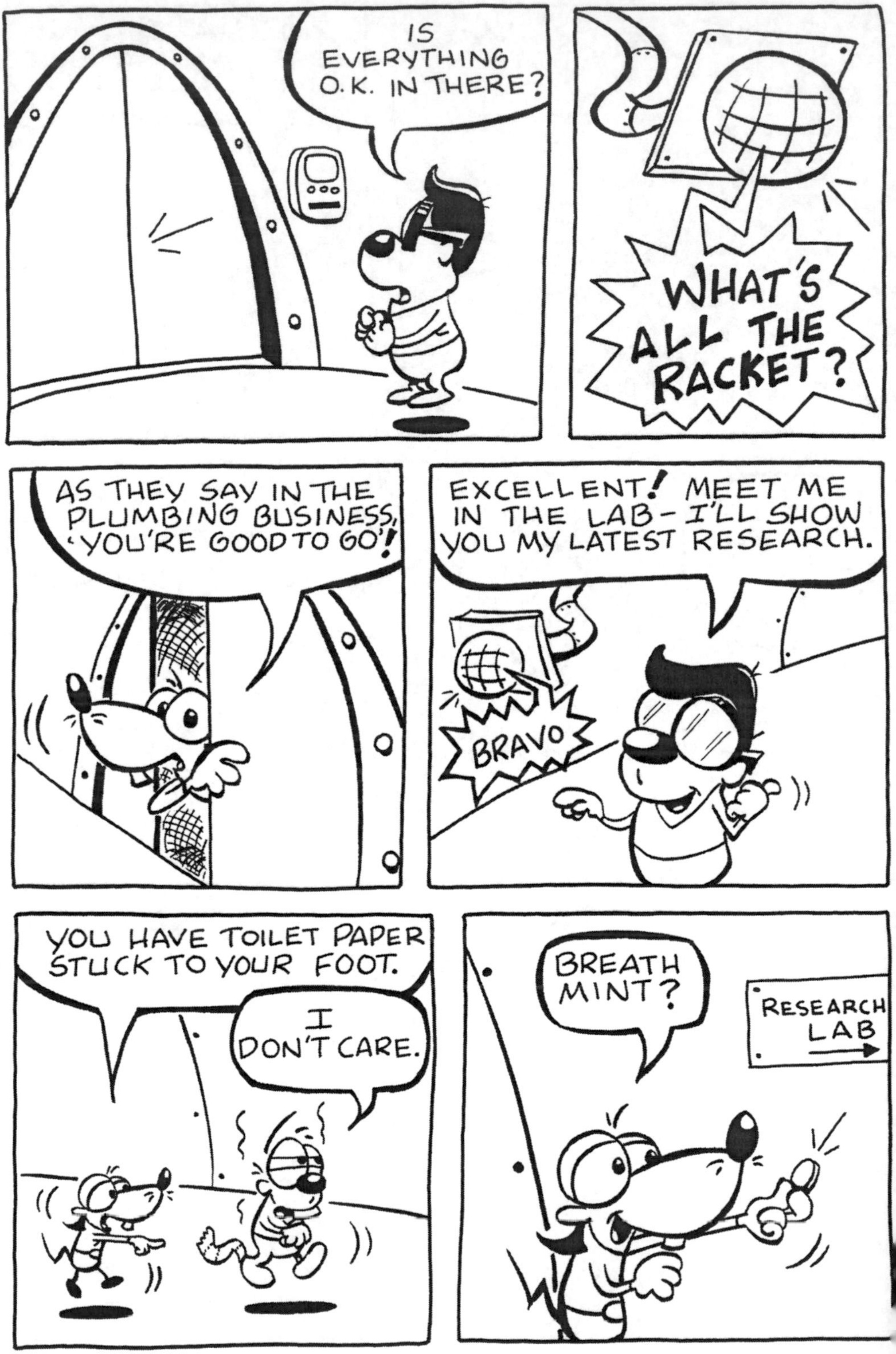
IS EVERYTHING O.K. IN THERE?
WHAT'S ALL THE RACKET?
AS THEY SAY IN THE PLUMBING BUSINESS, 'YOU'RE GOOD TO GO'!
EXCELLENT! MEET ME IN THE LAB - I'LL SHOW YOU MY LATEST RESEARCH.
BRAVO
YOU HAVE TOILET PAPER STUCK TO YOUR FOOT.
I DON'T CARE.
BREATH MINT?
RESEARCH LAB

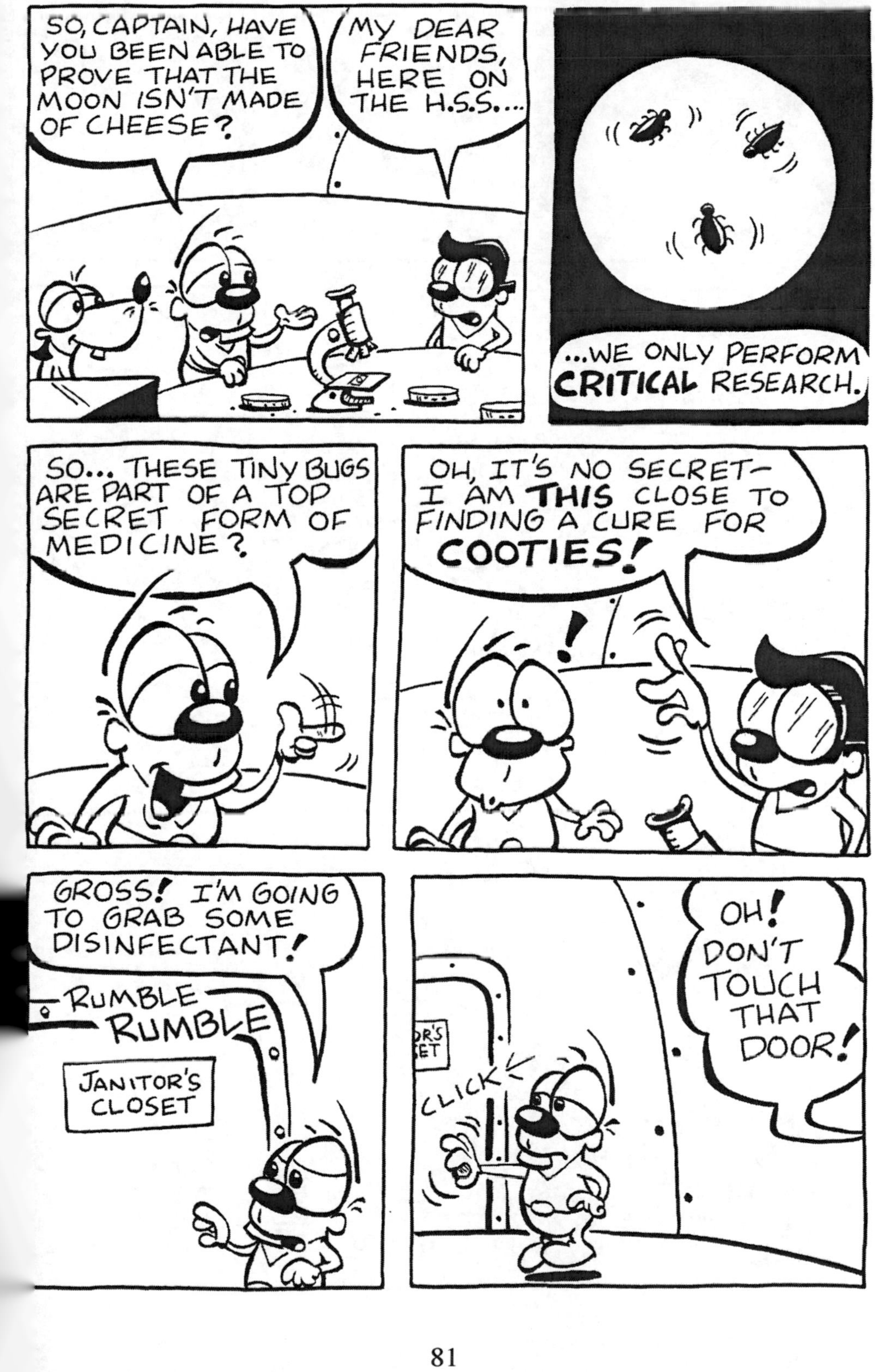
SO, CAPTAIN, HAVE YOU BEEN ABLE TO PROVE THAT THE MOON ISN'T MADE OF CHEESE?
MY DEAR FRIENDS, HERE ON THE H.S.S....
...WE ONLY PERFORM CRITICAL RESEARCH.
SO... THESE TINY BUGS ARE PART OF A TOP SECRET FORM OF MEDICINE?
OH, IT'S NO SECRET- I AM THIS CLOSE TO FINDING A CURE FOR COOTIES!
GROSS! I'M GOING TO GRAB SOME DISINFECTANT!
RUMBLE
RUMBLE
JANITOR'S CLOSET
CLICK
OH! DON'T TOUCH THAT DOOR!

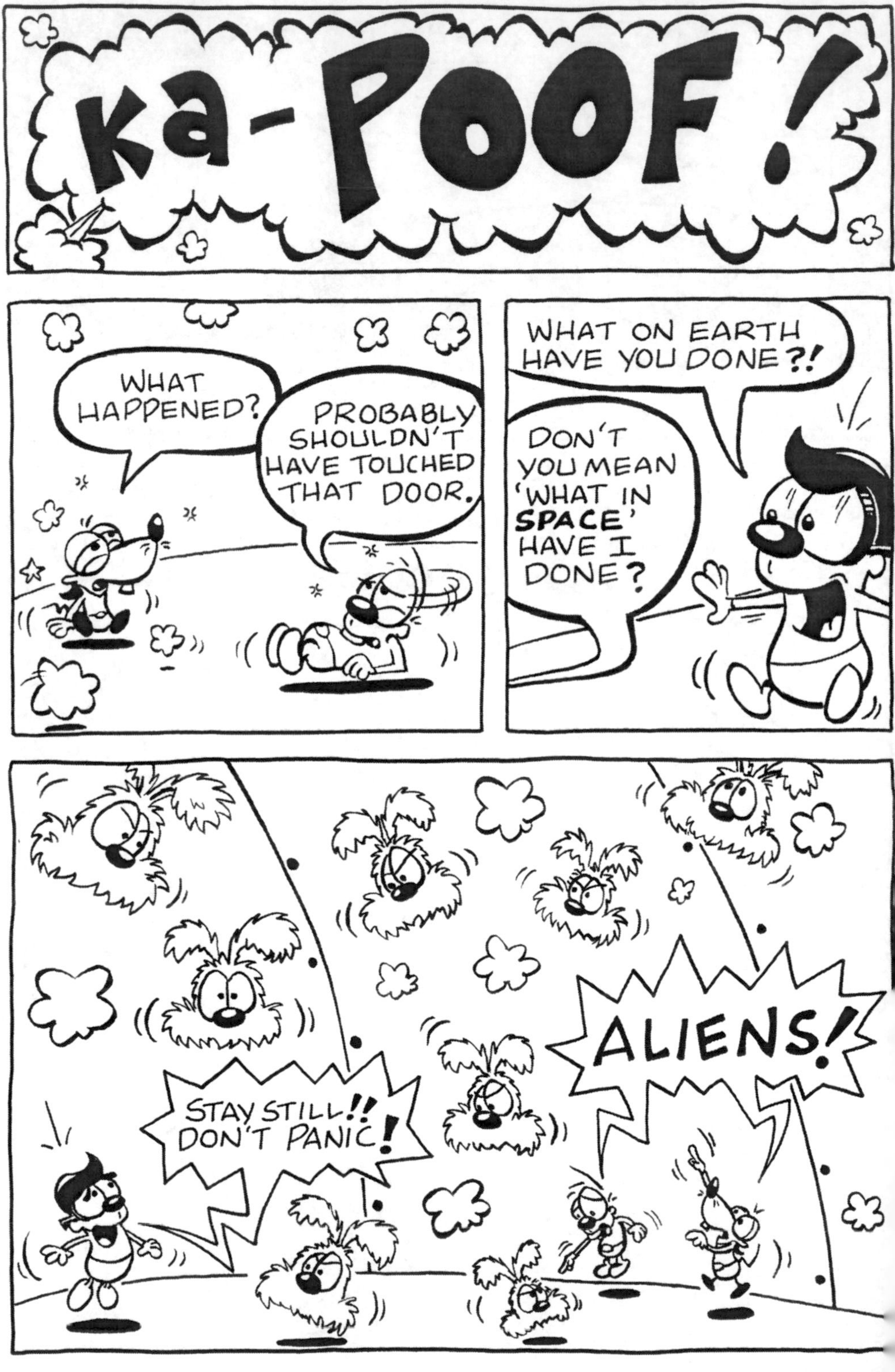
KA-POOF!
WHAT HAPPENED?
PROBABLY SHOULDN'T HAVE TOUCHED THAT DOOR.
WHAT ON EARTH HAVE YOU DONE?!
DON'T YOU MEAN 'WHAT IN SPACE' HAVE I DONE?
STAY STILL!! DON'T PANIC!
ALIENS!

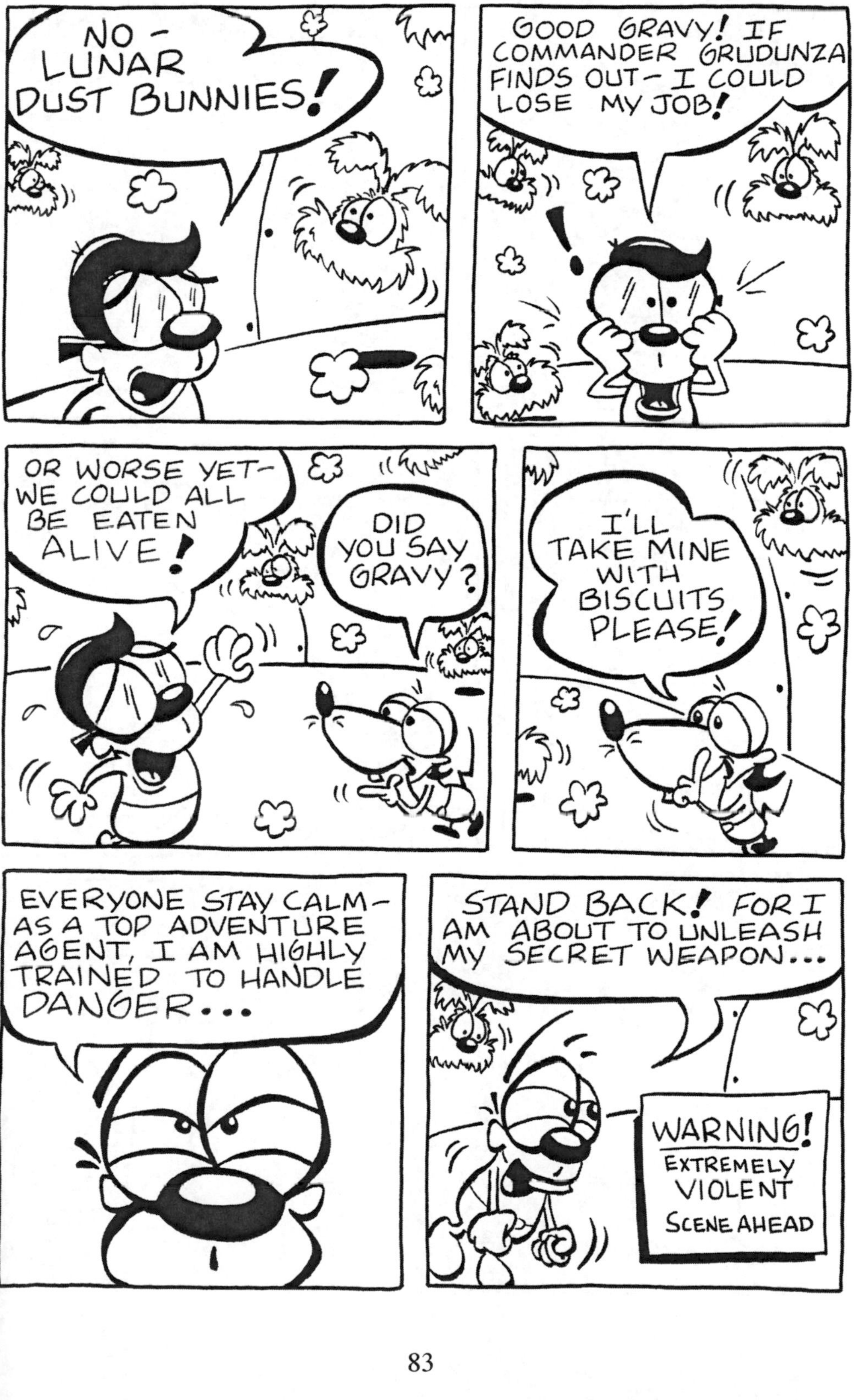
NO - LUNAR DUST BUNNIES!
GOOD GRAVY! IF COMMANDER GRUDUNZA FINDS OUT - I COULD LOSE MY JOB!
OR WORSE YET- WE COULD ALL BE EATEN ALIVE!
DID YOU SAY GRAVY?
I'LL TAKE MINE WITH BISCUITS PLEASE!
EVERYONE STAY CALM- AS A TOP ADVENTURE AGENT, I AM HIGHLY TRAINED TO HANDLE DANGER...
STAND BACK! FOR I AM ABOUT TO UNLEASH MY SECRET WEAPON...
WARNING! EXTREMELY VIOLENT SCENE AHEAD

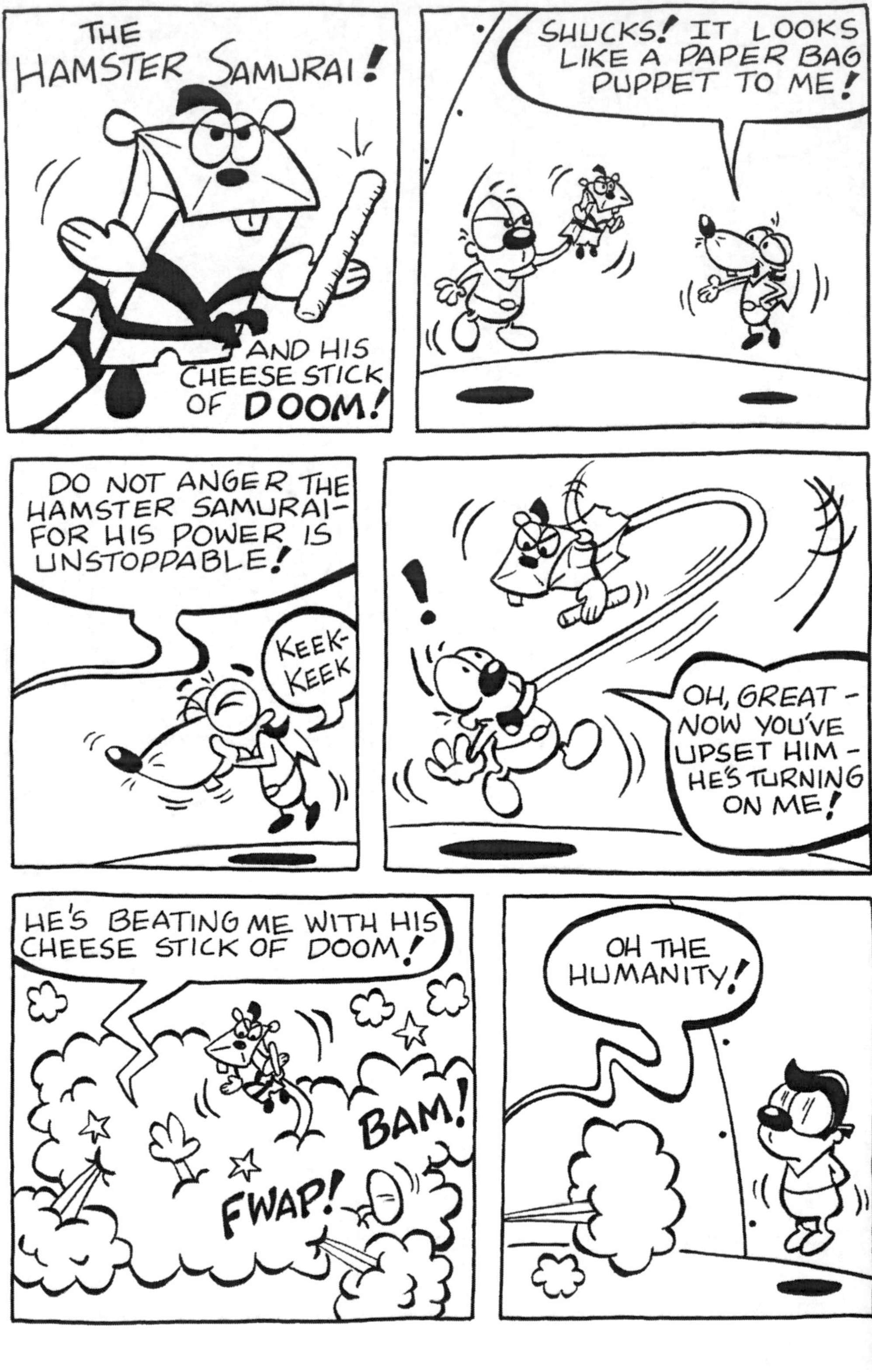
THE HAMSTER SAMURAI!
AND HIS CHEESE STICK OF DOOM!
SHUCKS! IT LOOKS LIKE A PAPER BAG PUPPET TO ME!
DO NOT ANGER THE HAMSTER SAMURAI- FOR HIS POWER IS UNSTOPPABLE!
KEEK-KEEK
OH, GREAT - NOW YOU'VE UPSET HIM - HE'S TURNING ON ME!
HE'S BEATING ME WITH HIS CHEESE STICK OF DOOM!
BAM!
FWAP!
OH THE HUMANITY!

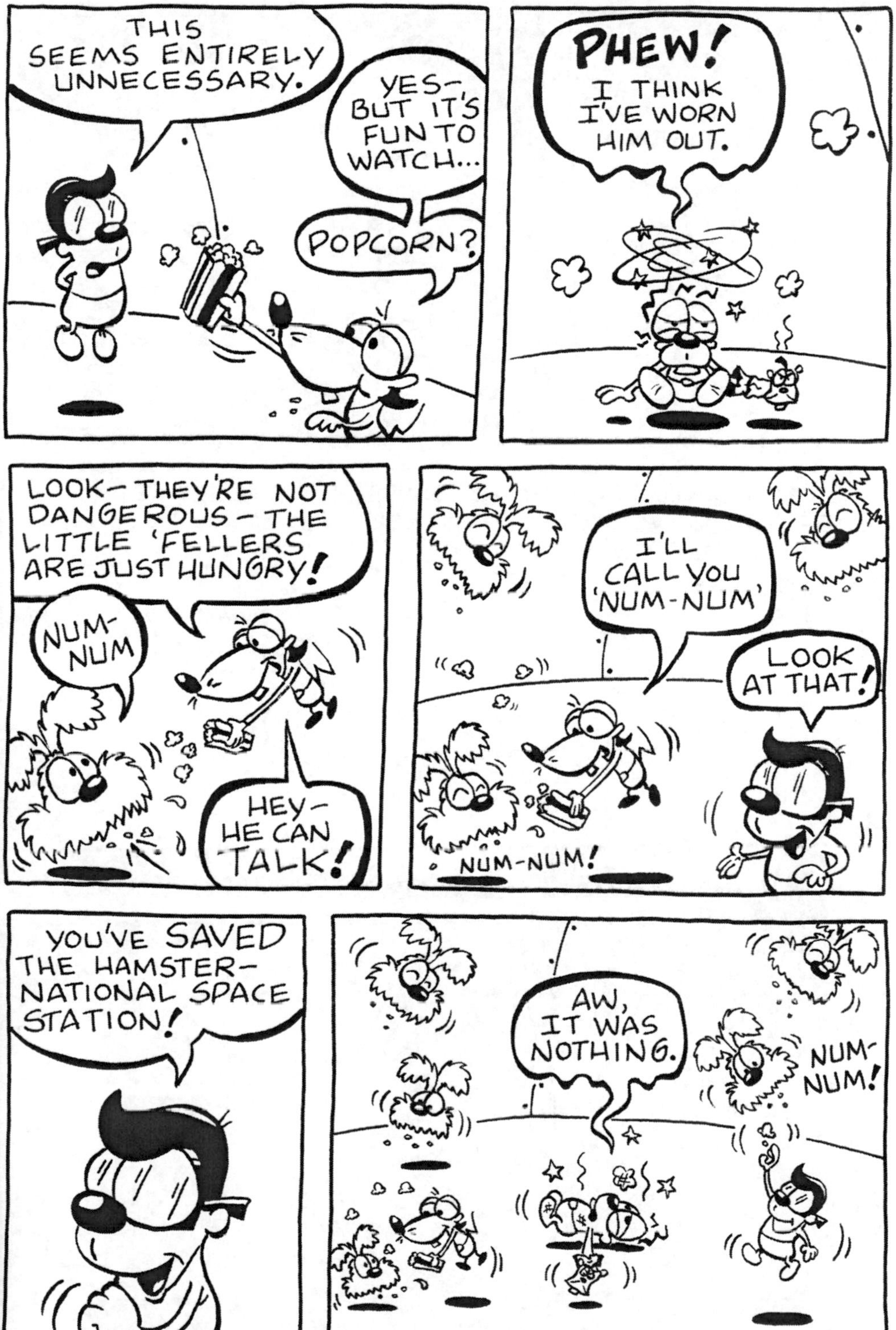

THIS SEEMS ENTIRELY UNNECESSARY.
YES— BUT IT'S FUN TO WATCH...
POPCORN?
PHEW! I THINK I'VE WORN HIM OUT.
LOOK— THEY'RE NOT DANGEROUS— THE LITTLE 'FELLERS ARE JUST HUNGRY!
NUM-NUM
HEY— HE CAN TALK!
I'LL CALL YOU 'NUM-NUM'
LOOK AT THAT!
NUM-NUM!
YOU'VE SAVED THE HAMSTER-NATIONAL SPACE STATION!
AW, IT WAS NOTHING.
NUM-NUM!

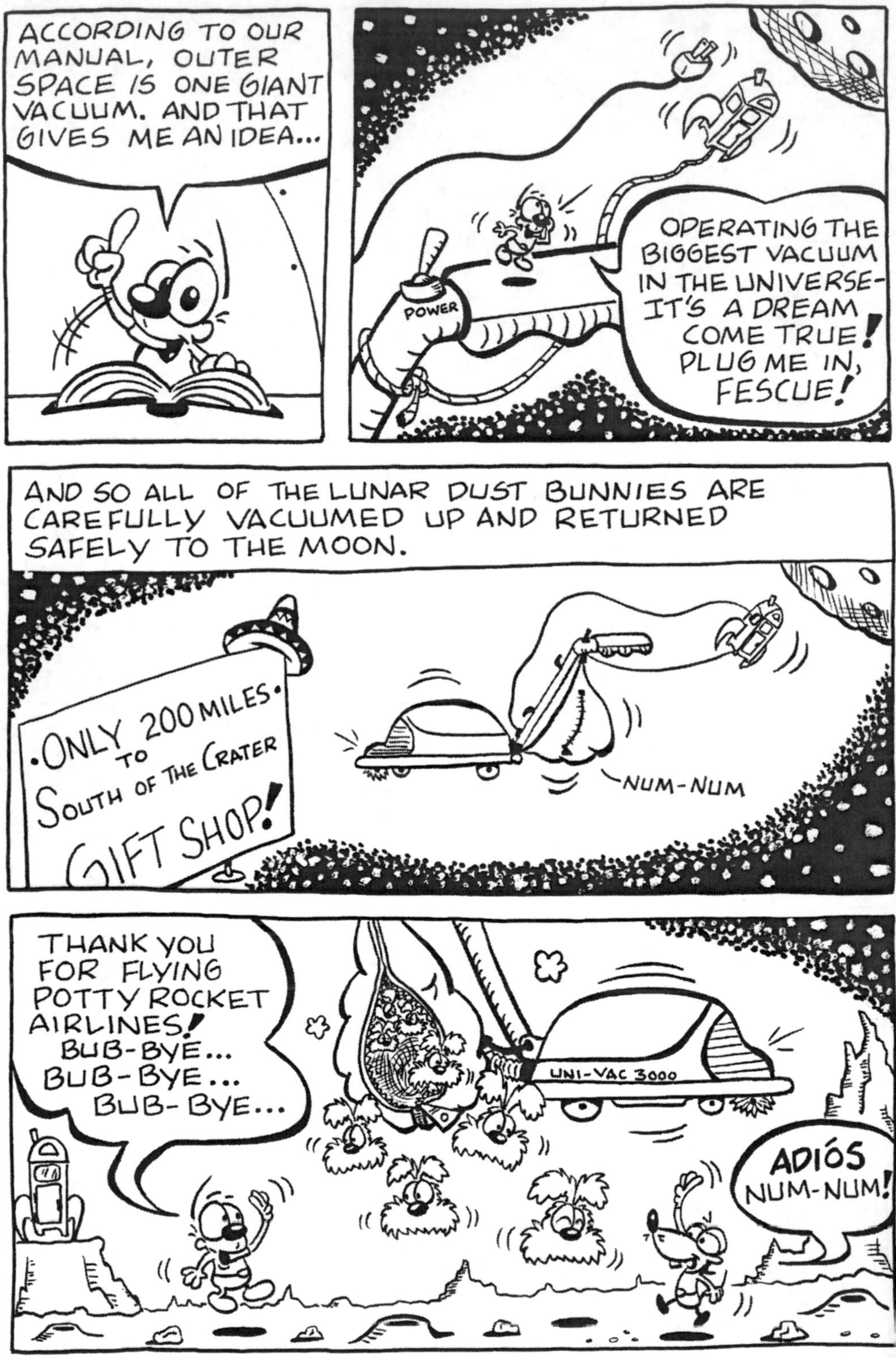
ACCORDING TO OUR MANUAL, OUTER SPACE IS ONE GIANT VACUUM. AND THAT GIVES ME AN IDEA...
POWER
OPERATING THE BIGGEST VACUUM IN THE UNIVERSE- IT'S A DREAM COME TRUE! PLUG ME IN, FESCUE!
AND SO ALL OF THE LUNAR DUST BUNNIES ARE CAREFULLY VACUUMED UP AND RETURNED SAFELY TO THE MOON.
•ONLY 200 MILES• TO SOUTH OF THE CRATER GIFT SHOP!
NUM-NUM
THANK YOU FOR FLYING POTTY ROCKET AIRLINES! BUB-BYE... BUB-BYE... BUB-BYE...
UNI-VAC 3000
ADIÓS NUM-NUM!

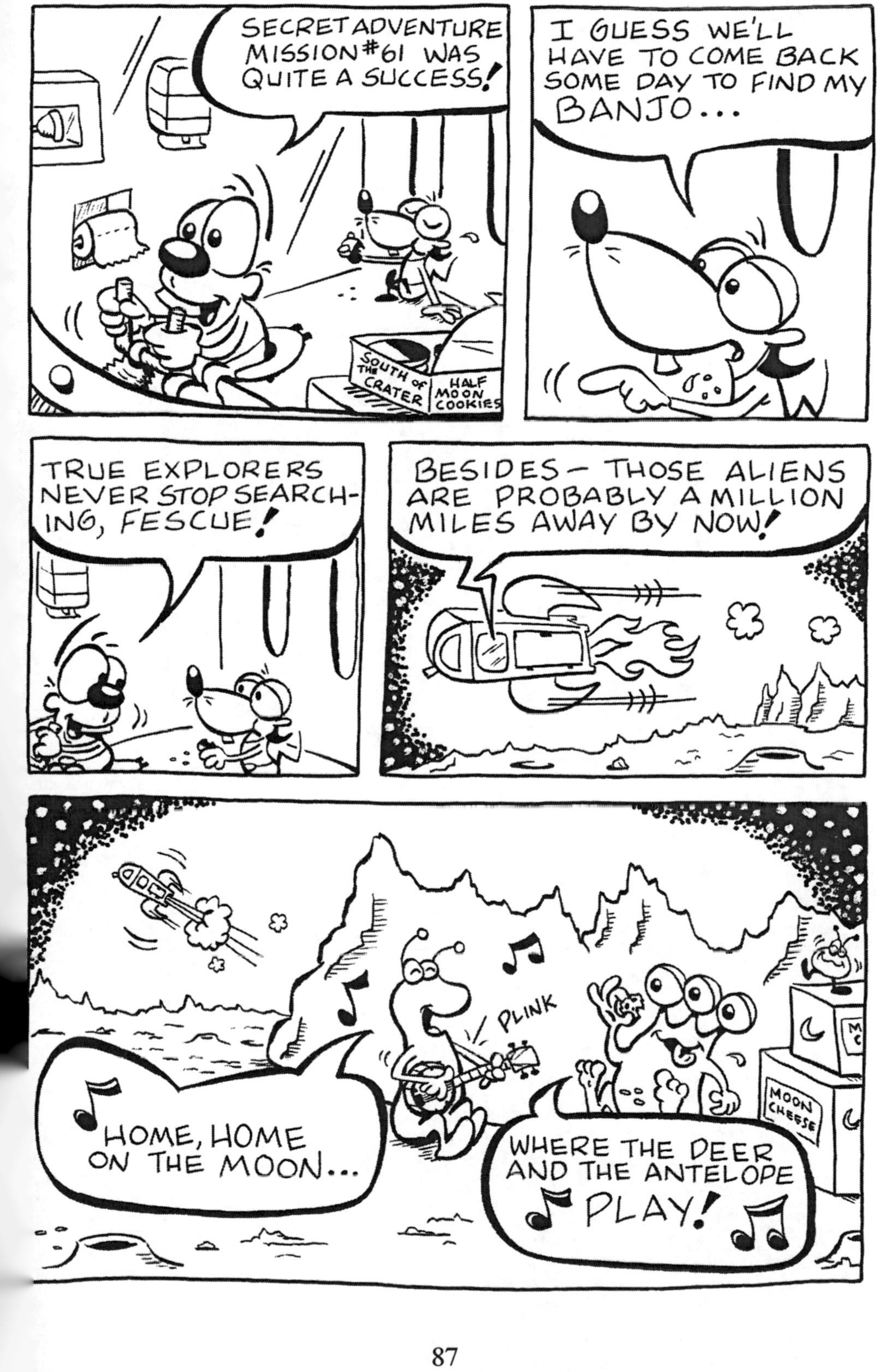
SECRET ADVENTURE MISSION #61 WAS QUITE A SUCCESS!
SOUTH OF THE CRATER
HALF MOON COOKIES
I GUESS WE'LL HAVE TO COME BACK SOME DAY TO FIND MY BANJO...
TRUE EXPLORERS NEVER STOP SEARCHING, FESCUE!
BESIDES— THOSE ALIENS ARE PROBABLY A MILLION MILES AWAY BY NOW!
PLINK
HOME, HOME ON THE MOON...
WHERE THE DEER AND THE ANTELOPE PLAY!
MOON CHEESE

Chapter Eight

Another Case Wrapped Up

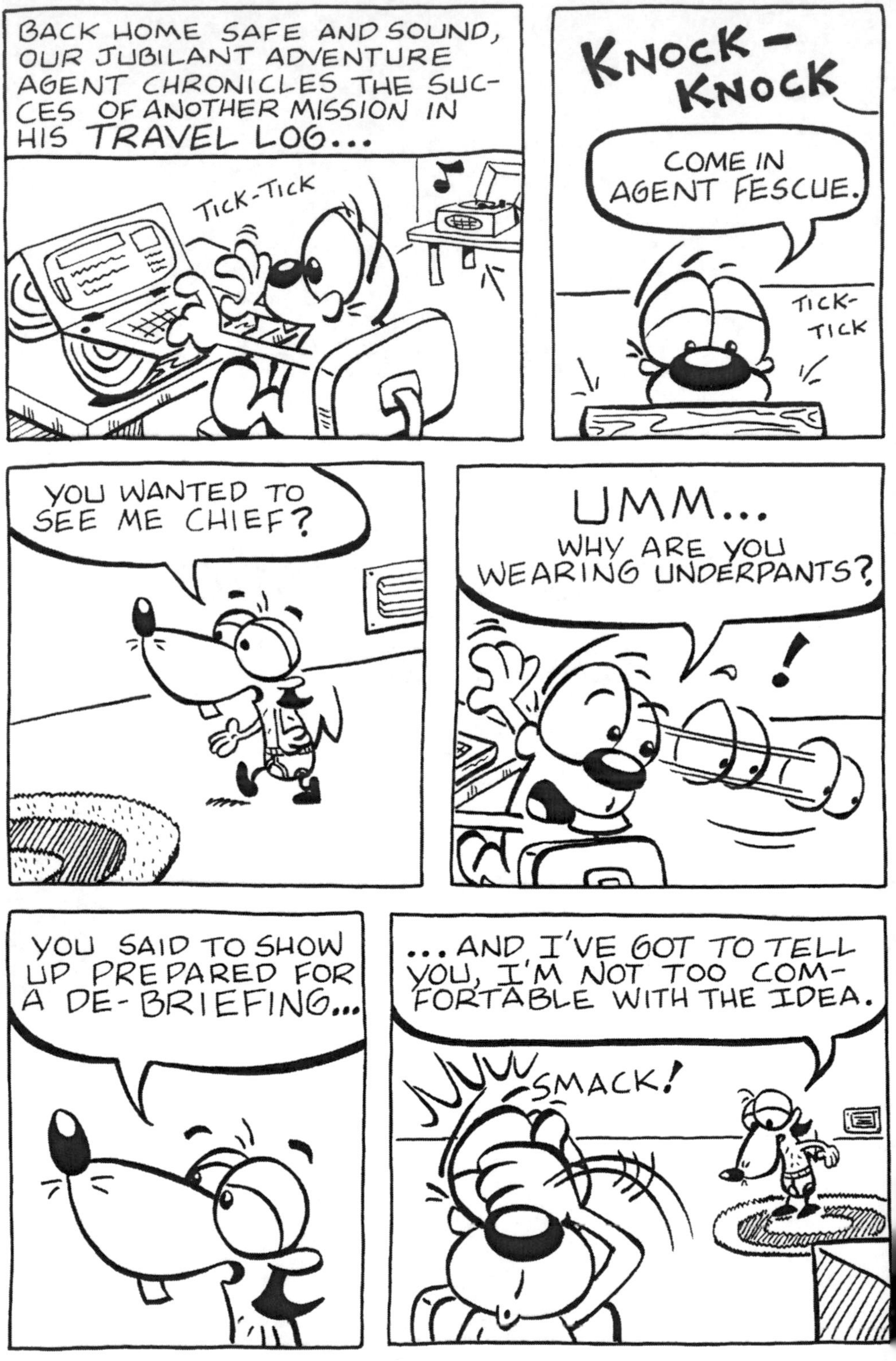
BACK HOME SAFE AND SOUND, OUR JUBILANT ADVENTURE AGENT CHRONICLES THE SUCCES OF ANOTHER MISSION IN HIS TRAVEL LOG...
TICK-TICK
KNOCK-KNOCK
COME IN AGENT FESCUE.
TICK-TICK
YOU WANTED TO SEE ME CHIEF?
UMM... WHY ARE YOU WEARING UNDERPANTS?
!
YOU SAID TO SHOW UP PREPARED FOR A DE-BRIEFING...
...AND I'VE GOT TO TELL YOU, I'M NOT TOO COMFORTABLE WITH THE IDEA.
SMACK!

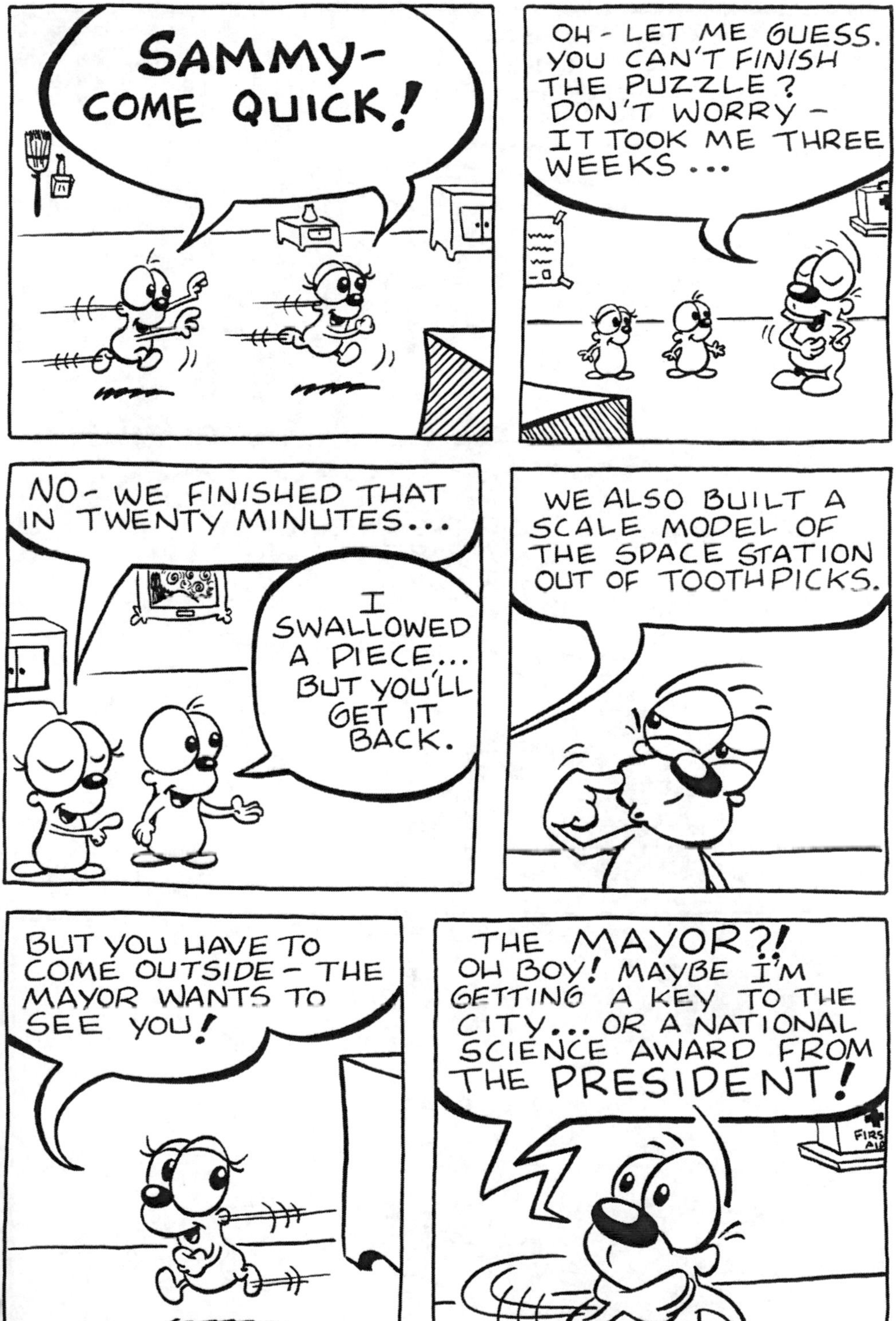
SAMMY- COME QUICK!
OH - LET ME GUESS. YOU CAN'T FINISH THE PUZZLE? DON'T WORRY - IT TOOK ME THREE WEEKS...
NO- WE FINISHED THAT IN TWENTY MINUTES...
I SWALLOWED A PIECE... BUT YOU'LL GET IT BACK.
WE ALSO BUILT A SCALE MODEL OF THE SPACE STATION OUT OF TOOTHPICKS.
BUT YOU HAVE TO COME OUTSIDE - THE MAYOR WANTS TO SEE YOU!
THE MAYOR?! OH BOY! MAYBE I'M GETTING A KEY TO THE CITY... OR A NATIONAL SCIENCE AWARD FROM THE PRESIDENT!

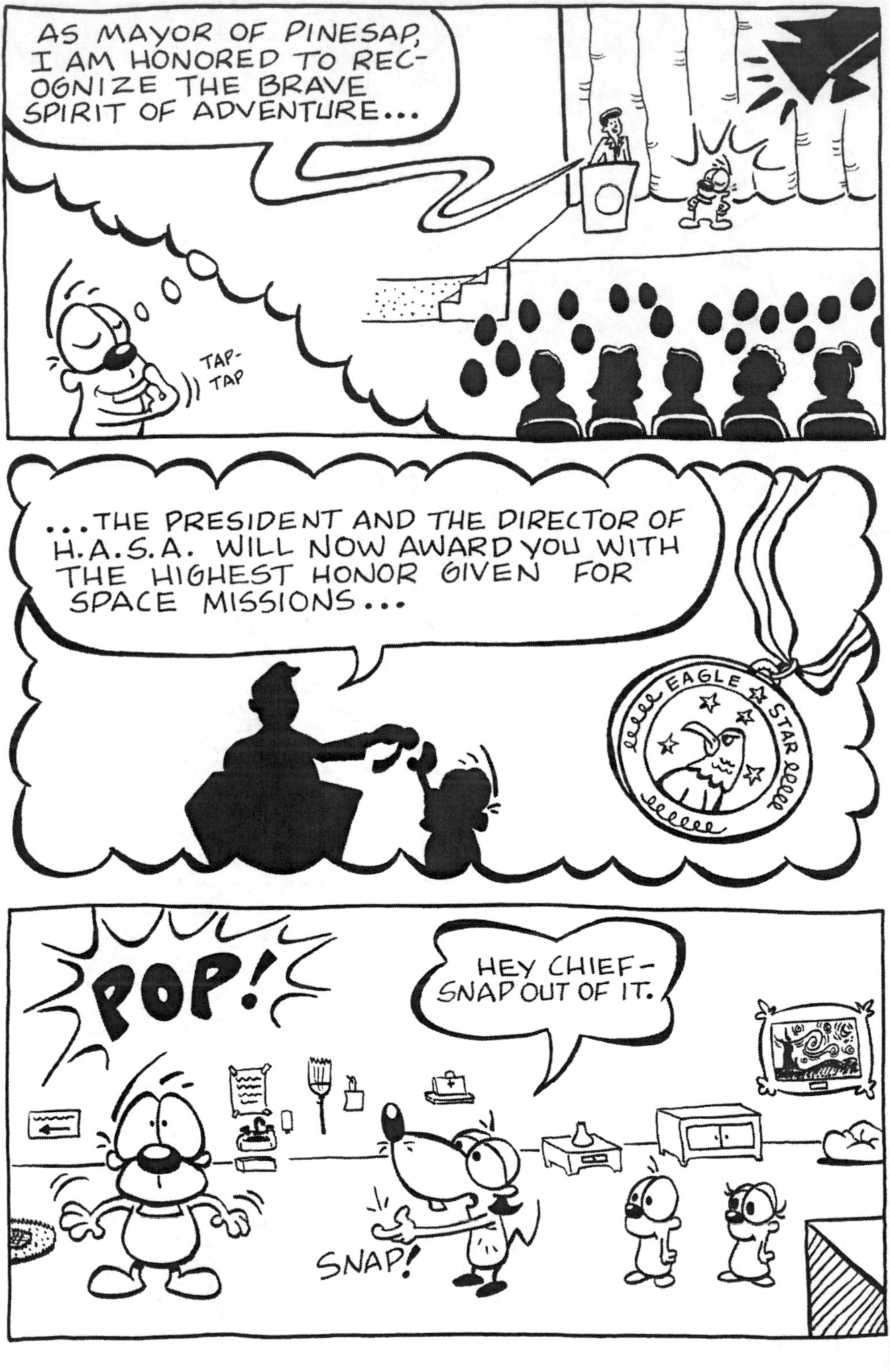
AS MAYOR OF PINESAP, I AM HONORED TO RECOGNIZE THE BRAVE SPIRIT OF ADVENTURE...
TAP-TAP
...THE PRESIDENT AND THE DIRECTOR OF H.A.S.A. WILL NOW AWARD YOU WITH THE HIGHEST HONOR GIVEN FOR SPACE MISSIONS...
EAGLE STAR
POP!
HEY CHIEF-SNAP OUT OF IT.
SNAP!

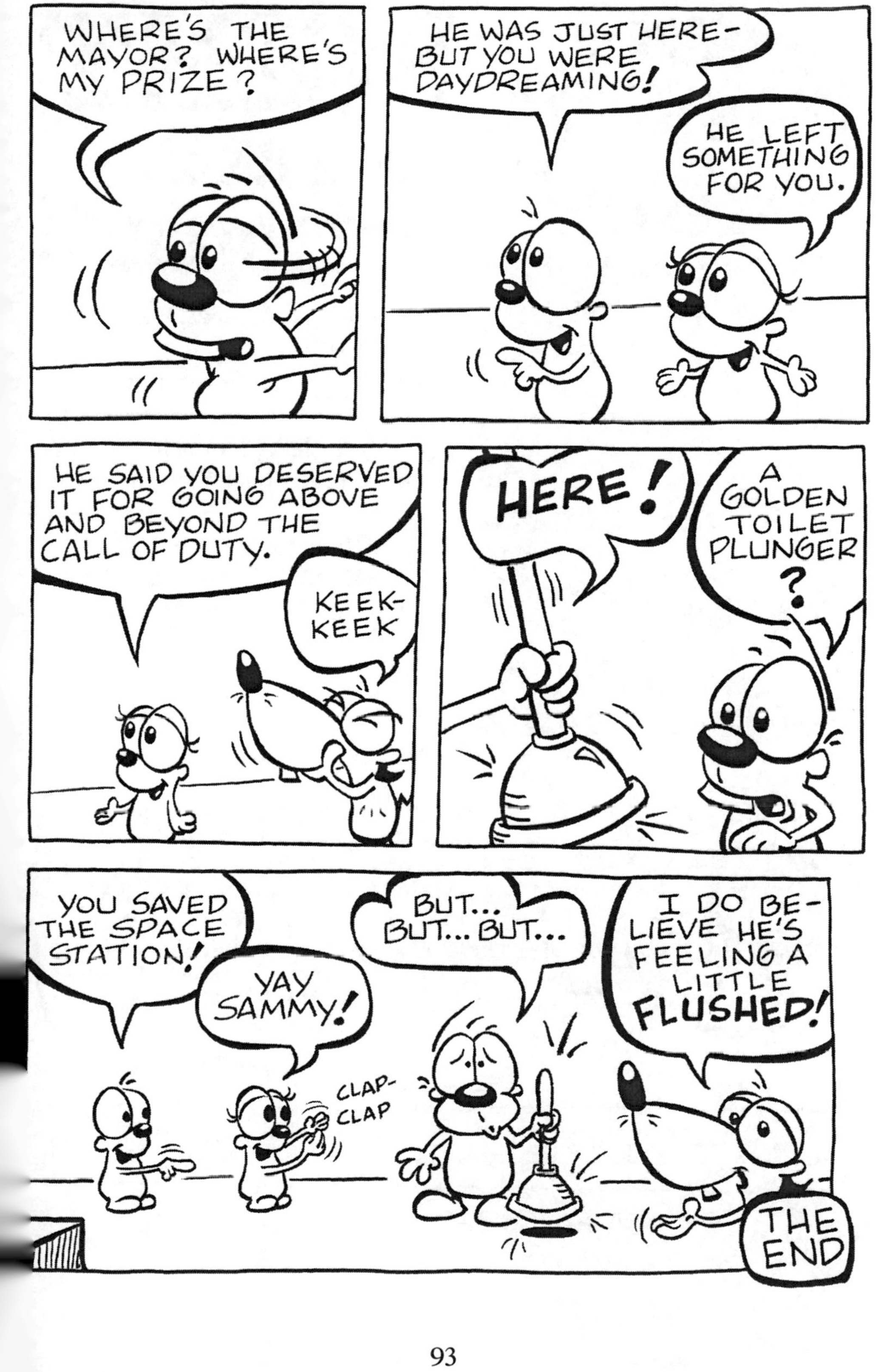
WHERE'S THE MAYOR? WHERE'S MY PRIZE?
HE WAS JUST HERE- BUT YOU WERE DAYDREAMING!
HE LEFT SOMETHING FOR YOU.
HE SAID YOU DESERVED IT FOR GOING ABOVE AND BEYOND THE CALL OF DUTY.
KEEK-KEEK
HERE!
A GOLDEN TOILET PLUNGER?
YOU SAVED THE SPACE STATION!
YAY SAMMY!
CLAP-CLAP
BUT... BUT... BUT...
I DO BE-LIEVE HE'S FEELING A LITTLE FLUSHED!
THE END

MEET
★ a REAL ★
Astronaut!

GREGORY H. JOHNSON
NASA Astronaut
Colonel, U.S. Air Force, Retired

Space Shuttle Missions to the
International Space Station:

Pilot, STS-123 in March, 2008
Pilot, STS-134 in October, 2011;
the final flight of the space shuttle
Endeavor.

GREETINGS, FELLOW SPACE LOVER! I WOULD LIKE TO SHARE A TOP-SECRET CONVERSATION I HAD RECENTLY WITH A REAL NASA ASTRONAUT. I CALL HIM 'ASTRONAUT GREG'. WHAT WOULD YOU ASK HIM IF YOU HAD THE CHANCE?

CLASSIFIED INFORMATION

FOR ADVENTURE SEEKERS ONLY!

"Astronaut Greg" flew in his T-38 supersonic jet and landed at the Pinesap International Airport to meet me. I'm still looking for his luggage! Pictured: Mr. McDonald (my cartoonist friend), "Astronaut Greg", me and Fescue.

Q & A

HAMSTER SAM: Is blasting off into space anything like riding a roller coaster?

ASTRONAUT GREG: The acceleration and vibration at liftoff is a little like the beginning of a roller coaster ride, when you crest the highest hill at the top and speed downward. But the abrupt turns on a coaster that cause side to side forces are more like flying a fighter aircraft than a spaceship.

HAMSTER SAM: From one pilot to another: have you ever had to pull over during a flight because your back seat passengers were arguing or fighting with one another?

ASTRONAUT GREG: Hahaha, that's a funny one, because I'm sure you could guess that has happened in my car – I'm a parent. It never happened in flight though, but I once cancelled a flight before takeoff because my wingman pilot was having a fight with his backseater. They were already out at their jets getting ready to go, but couldn't resolve their conflict without … the beginning indications of force. So, as the flight lead, I called off that particular flight before things got too physical … true story. Funny now – not so much at the time.

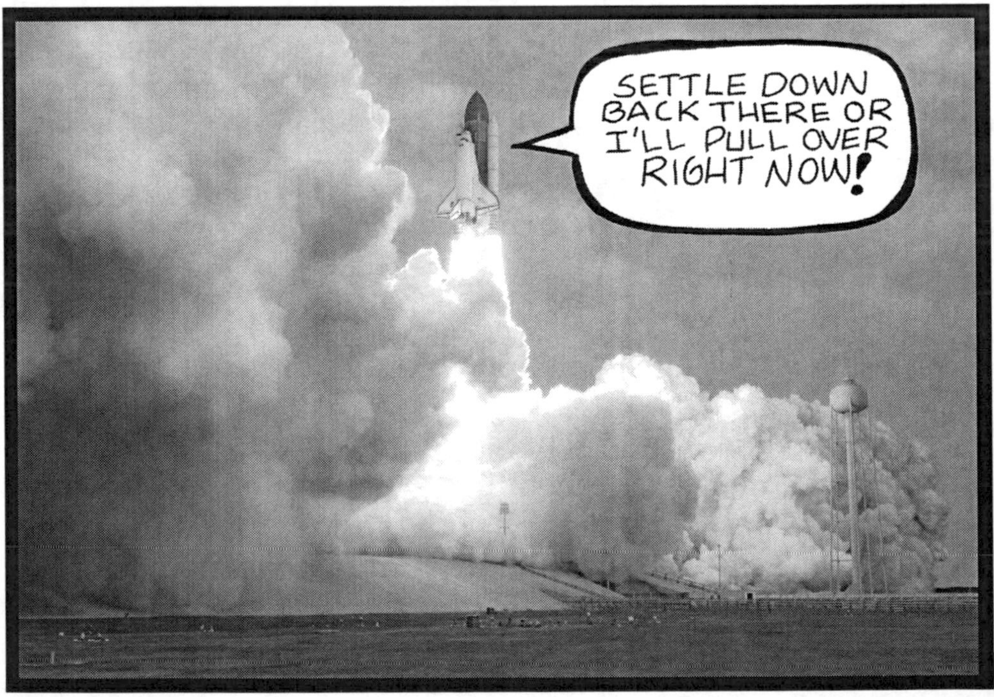
SETTLE DOWN BACK THERE OR I'LL PULL OVER RIGHT NOW!

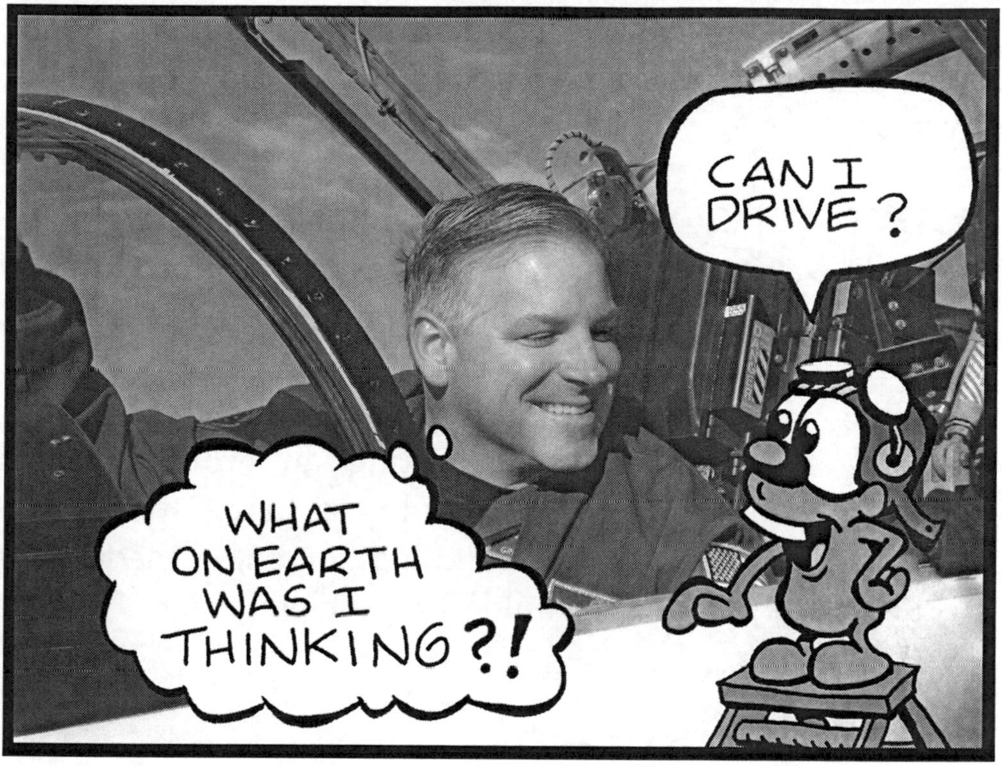
CAN I DRIVE?
WHAT ON EARTH WAS I THINKING?!

During our visit, I shared some of my own awesome piloting skills with "Astronaut Greg". He seemed to nod his head a lot and smile...

HAMSTER SAM: Do you think there may be any undiscovered planets in space? What will it take to discover new planets?

ASTRONAUT GREG: Yes, I am sure of that. The bigger question is: how many of them might be suitable to support life as we know it – or maybe some other form of life? We'll need a lot better rocket engines, so we can travel faster and farther – and faster ways to transmit information across vast distances.

FESCUE: Have you ever doodled in the margins of your important space mission papers? If so, did you get in trouble?

ASTRONAUT GREG: Doodling on our paper margins is the only way to keep track of changes to our procedures – so we do it all the time. So, the only time we get in trouble is if we doodle the wrong changes. Gotta be careful with doodling. If we're just doodling for fun, it's a good way to pass the time, especially during an orbit waveoff. What else can you do for 90 minutes buttoned up in your space suit? I guess if the doodling somehow made our papers unreadable, that would be a problem.

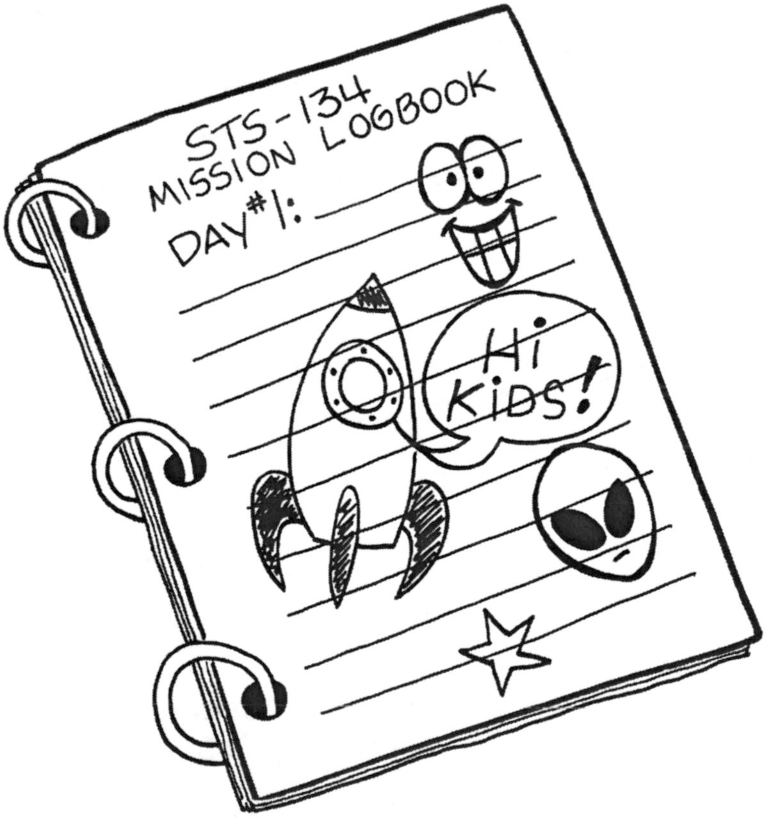
STS-134
MISSION LOGBOOK
DAY#1:
Hi KIDS!

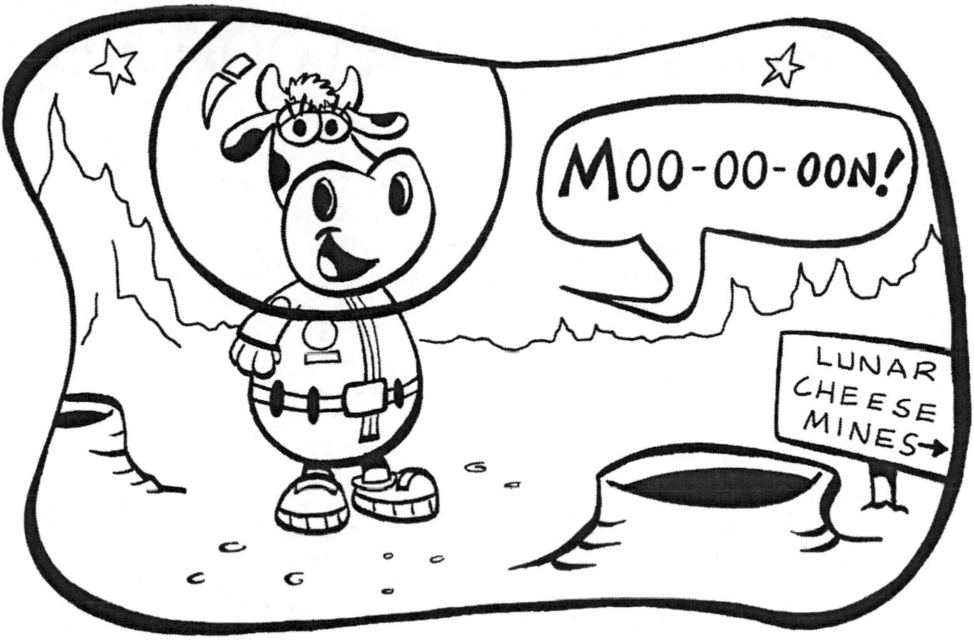
MOO-OO-OON!
LUNAR CHEESE MINES

HAMSTER SAM: Perhaps one day I shall return to outer space, at which time I will launch a mission to prove that, indeed, the moon is made of cheese! What say you, my fellow intergalactic explorer?

ASTRONAUT GREG: To be fair, Hamster Sam, we have only explored a small portion of the moon. Although we haven't found any cheese there to date, I can't 100% guarantee that there's no cheese hiding somewhere...but it's a pretty good bet that if you're a cheese lover, you should look elsewhere, like France or maybe Wisconsin. The moon is barren, dusty and lifeless. Not a great environment for cheese, at least the kind we have here on Earth!

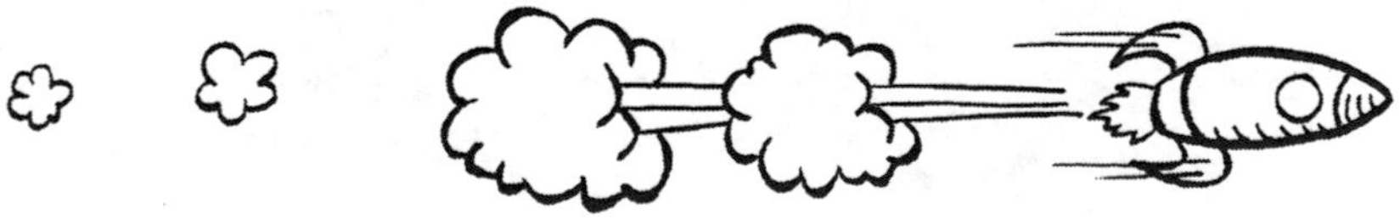

Explore Cartooning with Louise & Clark

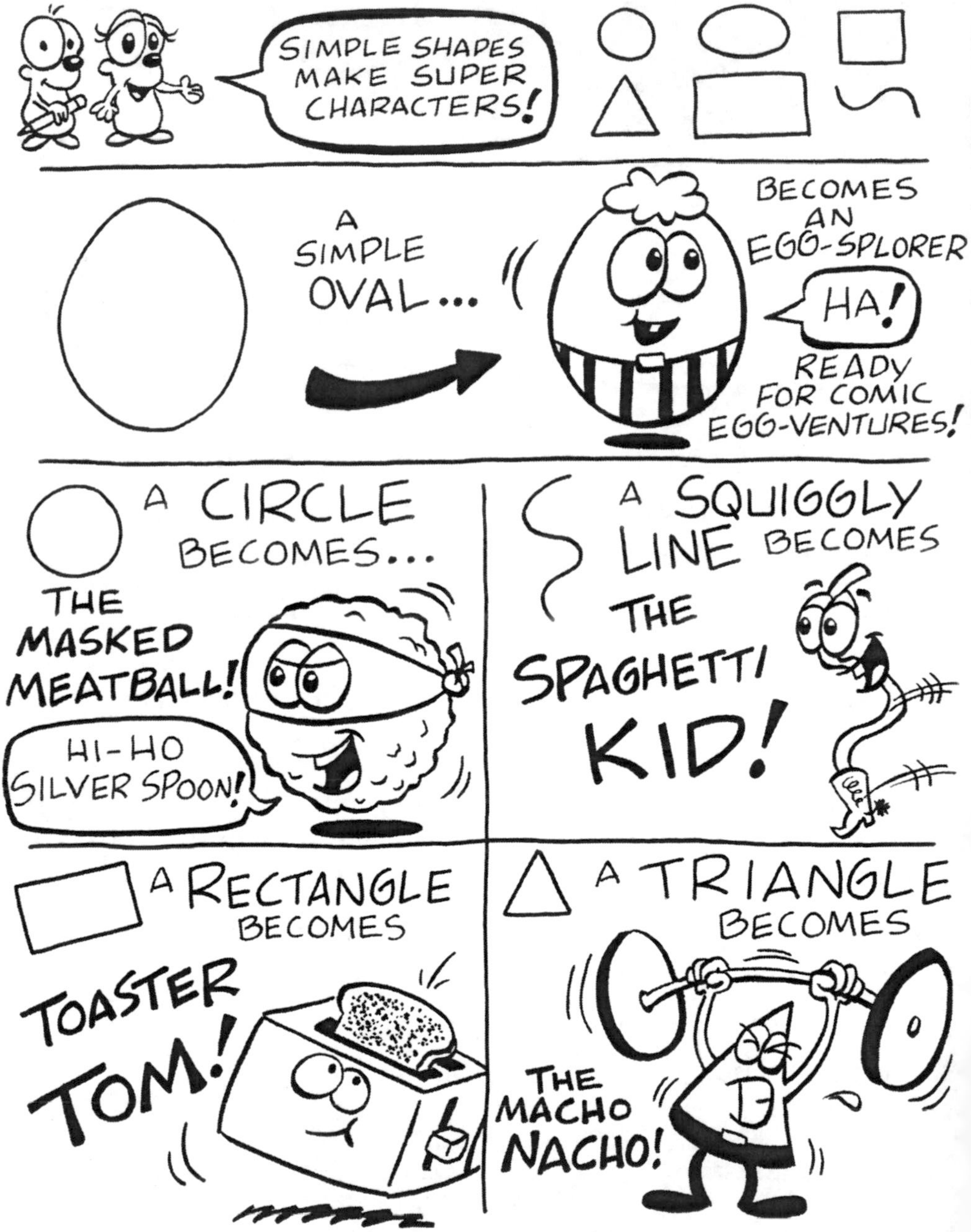
SIMPLE SHAPES MAKE SUPER CHARACTERS!
A SIMPLE OVAL...
BECOMES AN EGG-SPLORER
HA!
READY FOR COMIC EGG-VENTURES!
A CIRCLE BECOMES...
THE MASKED MEATBALL!
HI-HO SILVER SPOON!
A SQUIGGLY LINE BECOMES
THE SPAGHETTI KID!
A RECTANGLE BECOMES
TOASTER TOM!
A TRIANGLE BECOMES
THE MACHO NACHO!

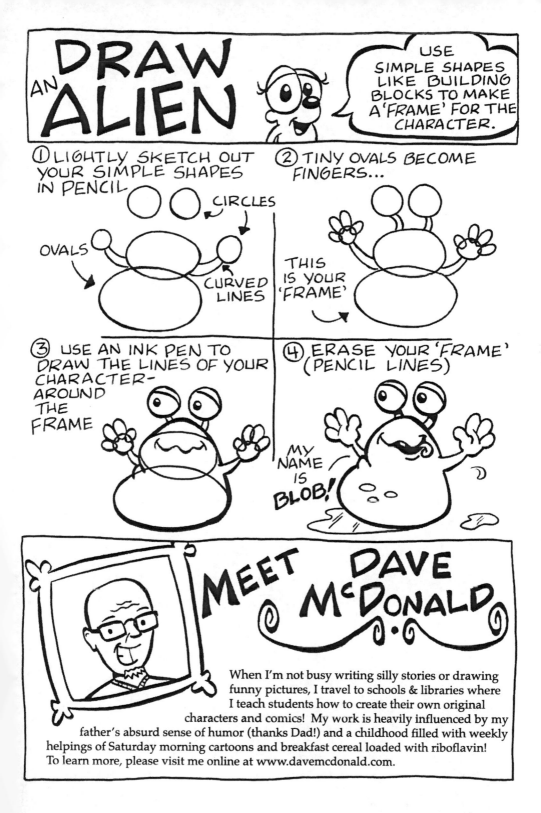

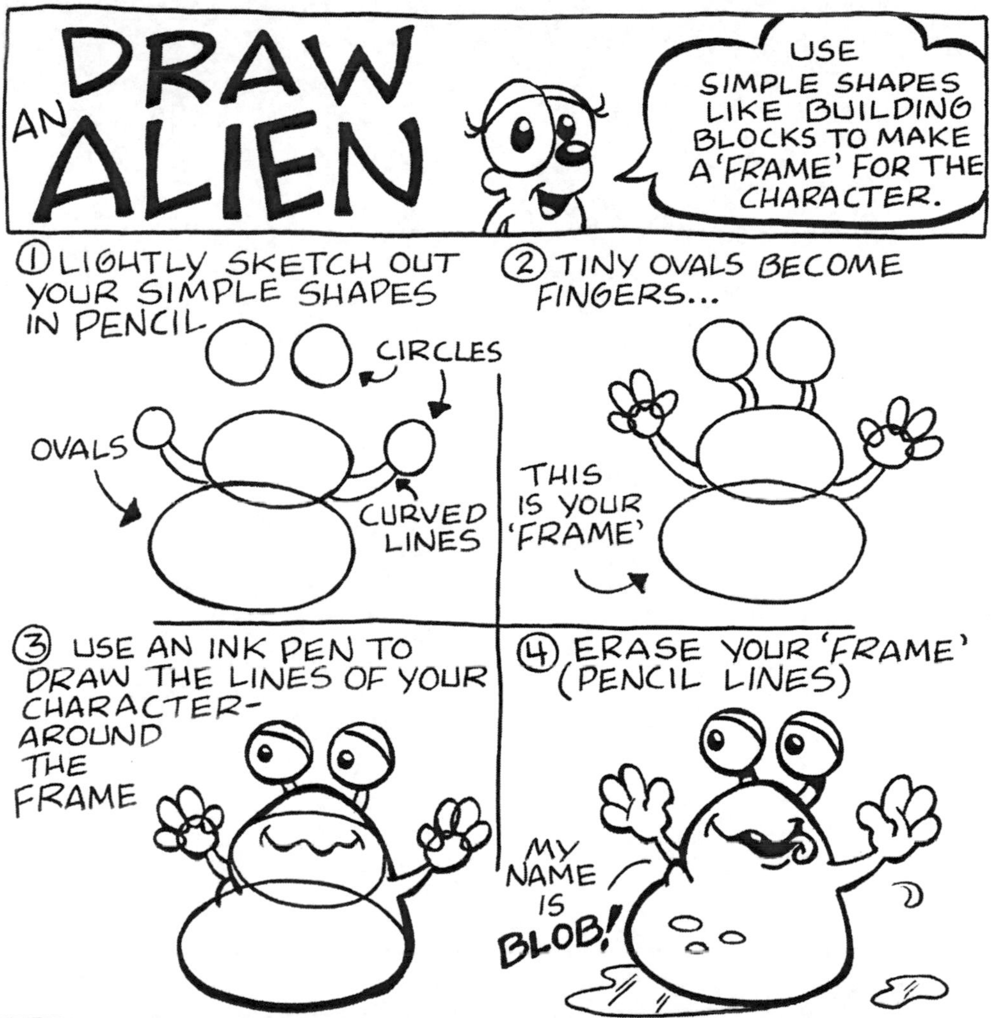
DRAW AN ALIEN
USE SIMPLE SHAPES LIKE BUILDING BLOCKS TO MAKE A 'FRAME' FOR THE CHARACTER.
① LIGHTLY SKETCH OUT YOUR SIMPLE SHAPES IN PENCIL
CIRCLES
OVALS
CURVED LINES
② TINY OVALS BECOME FINGERS...
THIS IS YOUR 'FRAME'
③ USE AN INK PEN TO DRAW THE LINES OF YOUR CHARACTER- AROUND THE FRAME
④ ERASE YOUR 'FRAME' (PENCIL LINES)
MY NAME IS BLOB!

MEET DAVE McDONALD

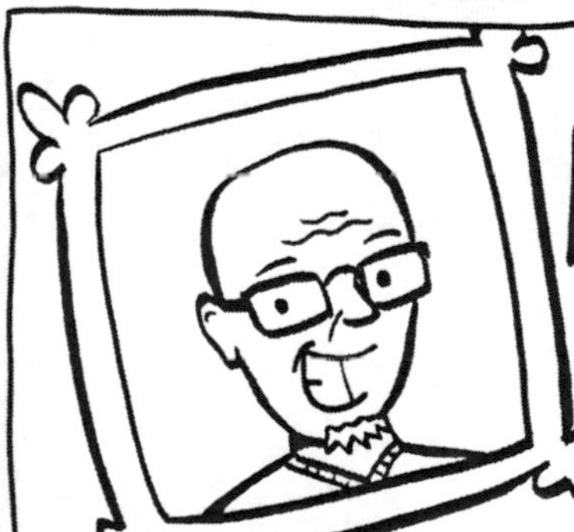

When I'm not busy writing silly stories or drawing funny pictures, I travel to schools & libraries where I teach students how to create their own original characters and comics! My work is heavily influenced by my father's absurd sense of humor (thanks Dad!) and a childhood filled with weekly helpings of Saturday morning cartoons and breakfast cereal loaded with riboflavin! To learn more, please visit me online at www.davemcdonald.com.

CPSIA information can be obtained at www.ICGtesting.com
Printed in the USA
LVOW10s1920270616

494289LV00008B/82/P

9 780979 844522